NAME ________________

AF691204

Round each number to the nearest hundred.

1. 247 ______ **2.** 671 ______ **3.** 364 ______

4. 892 ______ **5.** 2,732 ______ **6.** 4,091 ______

7. 6,280 ______ **8.** 7,709 ______ **9.** 8,079 ______

Round each decimal to the nearest whole number.

10. 0.9 ______ **11.** 4.3 ______ **12.** 5.7 ______

13. 23.4 ______ **14.** 0.74 ______ **15.** 3.19 ______

16. 14.62 ______ **17.** 70.26 ______ **18.** 39.04 ______

Write each improper fraction as a whole number or a mixed number.

19. $\frac{8}{4}$ ______ **20.** $\frac{12}{3}$ ______ **21.** $\frac{60}{20}$ ______

22. $\frac{35}{7}$ ______ **23.** $\frac{9}{5}$ ______ **24.** $\frac{7}{2}$ ______

Solve each problem.

25. Jill has $34.50. She wants to buy T-shirts that cost $4.99 each. How many can she buy?

______

26. Ted saved $65 for a new bike. The bike costs $98. How much more money does he need?

______

Use before pages 4–5.

NAME

Daily Review 2

Give at least two examples of different forms of numbers you would use in each situation. Name the form (whole number, fraction, decimal, or mixed number) of each number. (Need an example? See pages 4–5.)

1. Length of a car _____ ft _____ ft

Form of numbers: ____________________

2. Amount of rainfall in a day _____ in. _____ in.

Form of numbers: ____________________

3. Percentage of voters who voted for a particular candidate

_____ % _____ %

Form of numbers: ____________________

Choose +, −, ×, or ÷ for each. Then solve the problem.

4. Mr. Sanchez has 78 copies of a report to mail. The postage for each is $.44. How much will it cost him to mail the reports?

Operation: ____________

Answer: ____________

5. Mrs. Lewis has a piece of lumber that measures 45 in. She wants to cut it into 3 equal pieces. How long will each piece be?

Operation: ____________

Answer: ____________

NAME

Daily Review 3

Solve each problem. (Need an example? See pages 6–7.)

1. How many 3-digit numbers have a digital sum of 2?

____

2. I am a 3-digit number. All my digits are the same. My digital sum is 24. What am I?

____

3. What is the difference between the greatest 5-digit number and the least 5-digit number?

____

4. A number is formed by repeating the digits 123 four times. What is the 9th digit?

____

Give two examples of numbers you would use in each situation. Name the form (whole number, fraction, decimal, or mixed number) of each number. (Need an example? See pages 4–5.)

5. Length of caterpillar ______ in. ______ in.

Form of numbers: ______________________

6. Batting average of a baseball player ______________________

Form of numbers: ______________________

7. Wages per hour $______ $______

Form of numbers: ______________________

NAME

Daily Review 4

Choose paper and pencil or calculator to multiply. (Need an example? See pages 8–9.)

1. $0.5 \times 100 =$ ________

2. $24.39 \times 100 =$ ________

3. $0.06 \times 1{,}000 =$ ________

4. $8.73 \times 0.001 =$ ________

Use the table. Solve each problem. Round to the nearest cent.

Item	Price per pound	Item	Price per pound
apples	$0.56	onions	$0.34
chicken	$1.09	walnuts	$0.89

5. 1.8 lb of onions

____

6. 2.01 lb of walnuts

____

7. 4.6 lb of chicken

____

8. 7.92 lb of apples

____

Tell what each number means. (Need an example? See pages 4–5.)

9. 6:03 ________

10. $\frac{3''}{4}$ ________

11. $.69 ________

12. 7/4/92 ________

NAME

Daily Review 5

Estimate each product. Write the method of estimation.
(Need an example? See pages 10–13.)

1. 85×45 ____________

2. 23×81 ____________

3. 67×94 ____________

4. 36×48 ____________

Solve the problem. (Need an example? See pages 10–11.)

5. The cost of entrance into a school basketball game is $4.50. Children get in for half price. Estimate the cost for 3 adults and 3 children's tickets.

____

Choose paper and pencil or calculator to multiply. (Need an example? See pages 8–9.)

6. 457×1.2 ____________ **7.** 0.05×3.9 ____________

8. 28.51×0.06 ____________ **9.** 3.85×0.2 ____________

List the whole numbers from 1 to 40 in order to form one large number. (Need an example? See pages 6–7.)

1234567891011121314......3940

The digit "4" is in the 19th place. What digits are in the given places?

10. 21st ______ **11.** 45th ______ **12.** 63rd ______

NAME

Daily Review 6

Find each answer. Compute mentally.
(Need an example? See pages 16–17.)

1. $46 + 33$ ______
2. $419 + 328$ ______
3. 6×83 ______
4. $(53 \times 4) + (53 \times 6)$ ______
5. $762 - 178$ ______
6. $9{,}008 - 1{,}993$ ______

Estimate each answer. Write the method of estimation.
(Need an example? See pages 10–13.)

7. $2{,}523 + 3{,}590 + 4{,}889$ ______
8. $345 + 755$ ______
9. $4.6 + 8.1 + 9.3 + 5.2$ ______
10. \$23.99 + \$3.51 + \$62.75 ______
11. $432 - 198$ ______
12. 684×38 ______
13. 932×84 ______
14. $816 - 466$ ______

Solve the problem. (Need an example? See pages 6–7.)

15. How many 2-digit numbers have a digital sum of 6 and are odd numbers?

______

NAME

Daily Review 7

Divide. Choose paper and pencil or calculator.
(Need an example? See pages 18–21.)

1. $842 \div 7$ ______

2. $6{,}342 \div 5$ ______

3. $4\overline{)903}$

4. $3\overline{)2{,}285}$

5. $9\overline{)34{,}071}$

6. $6\overline{)19{,}470}$

Find each answer. Compute mentally.
(Need an example? See pages 16–17.)

7. $627 - 293$ ______

8. $(73 \times 6) + (73 \times 4)$ ______

9. $318 + 482$ ______

10. $65 + (894 + 35)$ ______

Choose paper and pencil or calculator to multiply.
(Need an example? See pages 8–9.)

11. $\begin{array}{r} 0.04 \\ \times\ \ 60 \\ \hline \end{array}$

12. $\begin{array}{r} 80 \\ \times 0.3 \\ \hline \end{array}$

13. $\begin{array}{r} 1.06 \\ \times\ \ 59 \\ \hline \end{array}$

14. $\begin{array}{r} 0.702 \\ \times\ 6.37 \\ \hline \end{array}$

15. $430 \times 8.5 =$ ______

16. $0.9 \times 0.3 =$ ______

17. $701.37 \times 0.002 =$ ______

18. $46.9 \times 5 =$ ______

Use before pages 22–23.

NAME

Daily Review 8

Use data from the graph to solve each problem. (Need an example? See pages 22–23.)

POPULATION OF THE U.S.

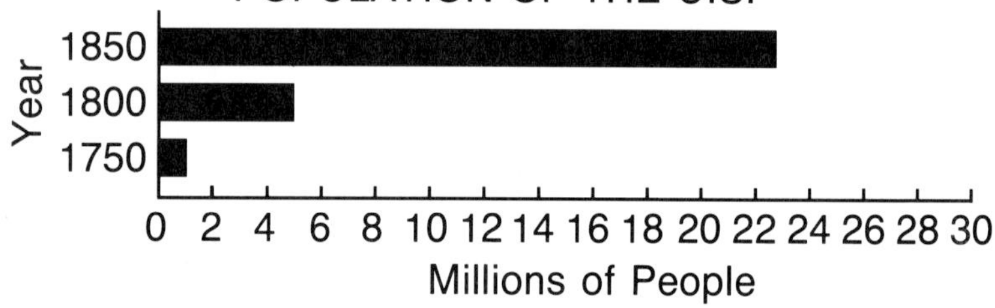

1. How much greater was the population in 1800 than in 1750?

   ____

2. What was the increase in population from 1750 to 1850?

   ____

3. About how many times greater was the population of 1850 than the population of 1800?

   ____

4. Was the increase in population from 1750 to 1800 greater than or less than that from 1800 to 1850?

   ____

Estimate each answer. Name the method you used.
(Need an example? See pages 10–14.)

5. 348 + 167 + 295 ______________________

6. 8,214 + 9,615 + 4,372 ______________________

7. 56.4 − 5.8 ______________________

8. 16 × 26 × 4 ______________________

NAME ______

Daily Review 9

Choose mental math, paper and pencil, or calculator to compute. Tell which method you chose.
(Need an example? See pages 24–25.)

1. 58×94 ______
2. 23×3 ______
3. $61 + 42$ ______
4. $5.089 + 0.68$ ______
5. $70 + 9$ ______
6. $791 - 326$ ______
7. $(6 + 7.2) \times 3.6$ ______
8. $268 + 902$ ______
9. $885 - 435$ ______

Divide. Choose paper and pencil or calculator.
(Need an example? See pages 18–21.)

10. $6\overline{)79}$ ______
11. $153 \div 7$ ______
12. $\frac{7,423}{4}$ ______
13. $\frac{38,252}{9}$ ______

Find each answer. Compute mentally.
(Need an example? See pages 16–17.)

14. $3,024 - 986$ ______
15. $56 + 194 + 37$ ______
16. 6×42 ______
17. $(73 \times 7) + (73 \times 3)$ ______

NAME

Daily Review 10

Write which operation to use. Then solve each problem.
(Need an example? See pages 26–27.)

1. A group of students were taken to a Nature Conservatory in 3 school buses and 6 minibuses. How many buses went on the trip in all?

Operation: ____________

Answer: ____________

2. At the conservatory, cows and horses graze on a field that measures 285 yd by 71 yd. What is the area of the field?

Operation: ____________

Answer: ____________

Choose mental math, paper and pencil, or calculator to compute. Write which method you chose.
(Need an example? See pages 24–25.)

3. $35{,}000 \div 5$ ____________ ____________

4. $249 + 767$ ____________ ____________

Divide. Choose paper and pencil or calculator. (Need an example? See pages 18–21.)

5. $4\overline{)926}$ **6.** $7\overline{)3{,}458}$ **7.** $6\overline{)709}$

8. $5\overline{)4{,}183}$ **9.** $9\overline{)18{,}604}$ **10.** $3\overline{)26{,}078}$

NAME

Daily Review 11

Complete the following tables. They describe plans for a crafts fair. (Need an example? See pages 38–41.)

1.

Children	Paint Jars
1	3
2	6
3	
	12
5	
6	

2.

Children	Paper
3	
4	
	25
	30
7	

Write which operation to use. Then solve each problem. (Need an example? See pages 26–27.)

3. Monday through Friday 785 people visited the fair. About how many visited on each day?

Operation: ______________

Answer: ______________

4. There were 72 local crafts people and 450 out-of-town crafts people at the fair. How many were there in all?

Operation: ______________

Answer: ______________

Choose mental math, paper and pencil, or calculator to compute. Write which method you choose and solve. (Need an example? See pages 24–25.)

5. 67×93 ______________ ______________

6. $16 + 21$ ______________ ______________

NAME ____________________

Daily Review 12

Evaluate $7x$ for each value of x. (Need an example? See pages 42–43.)

1. $x = 4$ ______ 2. $x = 0.1$ ______ 3. $x = 4.5$ ______

Evaluate $a - 8$ for each value of a.

4. $a = 13$ ______ 5. $a = 10.2$ ______ 6. $a = 11.6$ ______

Complete the following tables.
(Need an example? See pages 38–41.)

7.

m	n
0	0
1	4
2	
	12
	16

8.

g	h
1	1.7
2	2.7
	3.7
	4.7
5	

Write which operation to use. Then solve each problem.
(Need an example? See pages 26–27.)

9. A scientist has 7 packs of seeds. There are 14 seeds in a pack. How many seeds are there in all?

Operation: ____________

Answer: ____________

10. A farmer planted 90 early tomato plants and 60 late tomato plants. How many tomato plants were planted?

Operation: ____________

Answer: ____________

NAME

Daily Review 13

Write an expression for each phrase. (Need an example? See pages 44–45.)

1. 14 added to x ______

2. y subtracted from 7 ______

3. 7 subtracted from y ______

4. 28 divided by m ______

5. 35 added to c ______

6. 45 divided by p ______

7. The cost of r rubber balls that cost \$.85 each ______

8. The cost of p pens that cost \$.37 each ______

Evaluate $5 + b$ for each value of b. (Need an example? See pages 42–43.)

9. $b = 7$ ______ **10.** $b = 2.8$ ______ **11.** $b = 19$ ______

Evaluate $\frac{18}{d}$ for each value of d.

12. $d = 1$ ______ **13.** $d = 2$ ______ **14.** $d = 6$ ______

Complete the following tables.
(Need an example? See pages 38–41.)

15.

Children	10	11	12	13	
Paintbrushes	20	22	24		30

16.

Children	1	2	3		5
Liters of Water	1.5	3		6	

NAME

Daily Review 14

Solve each equation. (Need an example? See pages 46–47.)

1. $a - 20 = 75$ ________
2. $b + 22 = 40$ ________
3. $9 + e = 71$ ________
4. $38 = f - 12$ ________
5. $j - 0.4 = 9.1$ ________
6. $14 = k - 7.8$ ________

Write an expression for each phrase. (Need an example? See pages 44–45.)

7. 45 divided by x ________
8. w multiplied by 7 ________
9. 35 added to z ________
10. 15 subtracted from r ________
11. 120 divided by a ________
12. 9 subtracted from y ________

Evaluate $8a$ for each value of a.
(Need an example? See pages 42–43.)

13. $a = 1$ ________
14. $a = 5$ ________
15. $a = 4.6$ ________

Evaluate $b - 3$ for each value of b.

16. $b = 6$ ________
17. $b = 15$ ________
18. $b = 24$ ________

NAME

Daily Review 15

Solve each equation. (Need an example? See pages 48–49.)

1. $6b = 66$ ________
2. $3x = 45$ ________
3. $\frac{70}{x} = 10$ ________
4. $\frac{y}{15} = 3$ ________
5. $6m = 18$ ________
6. $7s = 49$ ________
7. $\frac{a}{8} = 3$ ________
8. $\frac{81}{c} = 9$ ________
9. $3p = 78$ ________
10. $8k = 12$ ________

Write and solve an equation.
(Need an example? See pages 46–47.)

11. A bus with a capacity of 42 seated passengers now has 37 passengers aboard. How many empty seats are there? ________

12. Ron needs 312 eggs to fill the order from the Newtown supermarket. He now has 255 eggs. How many more eggs does he need? ________

Write an expression for each of the following. (Need an example? See pages 44–45.)

13. The cost of q jars of paint that cost \$.95 each ________
14. The cost of r colored pencils that cost \$.53 each ________
15. The cost of p books that cost \$4.95 each ________
16. The cost of s combs that cost \$.28 each ________

NAME

Daily Review 16

Write an equation that fits the given information about a number of swimmers, *s*, on the coast last Saturday. Then solve the equation. (Need an example? See pages 52 and 53.)

1. If there were 11 more swimmers at Rocky Beach, there would have been 600 swimmers there.

____

2. If there were 19 fewer swimmers at Stony Beach, there would have been 600 swimmers there.

____

Solve each equation. Remember to look at the operation sign. (Need an example? See pages 48–49.)

3. $6m = 60$ ________

4. $4n = 60$ ________

5. $34 = 2v$ ________

6. $\frac{r}{3} = 34$ ________

7. $\frac{s}{100} = 300$ ________

8. $\frac{t}{20} = 600$ ________

Solve each equation. (Need an example? See pages 46–47.)

9. $a + 15 = 35$ ________

10. $b - 49 = 72$ ________

11. $0.9 = f + 0.8$ ________

12. $4.1 + d = 8.3$ ________

13. $e - 35.7 = 70.2$ ________

NAME

Daily Review 17

Compute. (Need an example? See pages 54–55.)

1. $\frac{(14 - 4)}{2}$ ______ **2.** $14 - \frac{14}{2}$ ______

3. $\frac{14}{2} - 4$ ______ **4.** $6 + 3 \times 8 + 2$ ______

5. $5 \times 6 - 7 \times 4$ ______ **6.** $\frac{15}{3} + 2$ ______

7. $\frac{11 - 2 + 9}{9}$ ______ **8.** $\frac{7 + 3}{2} + 12$ ______

Write an equation that fits the given information about the number of specimens of rock, *r,* in the natural history museum. Then solve each equation. (Need an example? See pages 52–53.)

9. If there were 55 more specimens, there would have been 500 specimens in the museum.

______

10. Half of the specimens are in The Hull Wing of the museum. There are 672 specimens in this wing.

______

Solve each equation. (Need an example? See pages 48–49.)

11. $4x = 36$ ______ **12.** $\frac{y}{7} = 3$ ______ **13.** $5z = 70$ ______

14. $\frac{p}{6} = 4$ ______ **15.** $8r = 64$ ______ **16.** $\frac{25}{t} = 5$ ______

NAME ______________________

Daily Review 18

Evaluate $4 + 3a$ for each value of a. (Need an example? See pages 56–57.)

1. $a = 1$ ______ **2.** $a = 2$ ______ **3.** $a = 3$ ______

Evaluate $\frac{15}{b} + b$ for each value of b.

4. $b = 1$ ______ **5.** $b = 3$ ______ **6.** $b = 5$ ______

Compute. (Need an example? See pages 54–55.)

7. $\frac{15}{(3 + 2)}$ ______ **8.** $\frac{42}{6} \times 7$ ______

9. $\frac{2 \times (4 + 6)}{5}$ ______ **10.** $\frac{7 \times 7 - 2 \times 2}{9 + 6}$ ______

11. $\frac{\frac{8}{2} + 5}{5 - 2}$ ______ **12.** $\frac{\frac{44}{4} + 3}{9 - 2}$ ______

Write an equation that fits the given information about the number of members, m, there are on a chess team. Then solve each equation. (Need an example? See pages 52–53.)

13. Half the number of members is 9.

____

14. Four times the number of members is 32.

____

15. If there were 7 fewer members on the team, the total number of team members would be 29.

____

16. The number of members tripled is 42.

____

NAME

Daily Review 19

Write an expression for each of the following. (Need an example? See pages 58–59.)

1. 3 less than a ________ **2.** 9 times b ________

3. the sum of c and d ________

4. the product of 8 and g ________

5. 4 more than k divided by 3 ________

6. 2 less than p divided by 4 ________

Complete each table. (Need an example? See pages 56–57.)

7.

g	$1.4 + 3g$
1	
2	
3	
4	
10	

8.

h	$\frac{h}{5} + 9$
10	
20	
30	
40	
100	

Evaluate each expression when $c = 12$.

9. $2c - 4$ ____ **10.** $6c - 12$ ____ **11.** $\frac{(c + 6)}{3}$ ____

Evaluate each expression when $p = 7$.

12. $\frac{(p - 3)}{2}$ ____ **13.** $5p - 15$ ____ **14.** $3p + 6$ ____

  Use before pages 60–61.

NAME ____________________

Daily Review 20

Solve each equation. (Need an example? See pages 60–61.)

1. $3m + 19 = 40$ ________ **2.** $2n - 11 = 53$ ________

3. $\frac{p}{4} - 5 = 3$ ________ **4.** $\frac{g}{5} + 4 = 14$ ________

5. $5r + 2.3 = 12.3$ ________ **6.** $3s + 14.92 = 20.92$ ________

7. $\frac{t}{7} + 2 = 16$ ________ **8.** $\frac{v}{9} - 18 = 4$ ________

The distance an object has gone can be computed from its rate (or speed) and the time it has traveled. Write an expression for each of the following. Use r and t. (Need an example? See pages 58–59.)

9. rate times time ____________________

10. time divided by rate ____________________

11. rate divided by time ____________________

12. rate times twice the time ____________________

13. three times the rate times four times the time ________

Evaluate $6 + 3x$ for each value of x. (Need an example? See pages 56–57.)

14. $x = 1$ ______ **15.** $x = 2$ ______ **16.** $x = 3$ ______

17. $x = 7$ ______ **18.** $x = 9$ ______ **19.** $x = 5$ ______

NAME

Daily Review 21

Use the formula $P = \frac{d}{33} + 1$ to solve these problems. (Need an example? See pages 62–63.)

How many atmospheres of pressure would there be at each depth?

1. 132 feet ____________

2. 165 feet ____________

3. 396 feet ____________

4. 231 feet ____________

Find the depth (in feet) for each of the following amounts of pressure.

5. 10 atmospheres ____________

6. 4 atmospheres ____________

Solve each equation. (Need an example? See pages 60–61.)

7. $6a + 6 = 24$ ____________

8. $2b - 3.21 = 3.21$ ____________

9. $8.5 + 4c = 24.5$ ____________

10. $\frac{d}{9} + 17 = 62$ ____________

11. $\frac{a}{7} + 8 = 12$ ____________

12. $4f - 16.48 = 23.52$ ____________

NAME

Daily Review 22

Divide using mental math. (Need an example? See pages 74–75.)

1. $5{,}600 \div 70$ ________
2. $300 \div 50$ ________
3. $280{,}000 \div 400$ ________
4. $180{,}000 \div 600$ ________
5. $\frac{350{,}000}{700}$ ________
6. $\frac{360{,}000}{9{,}000}$ ________
7. $\frac{600{,}000}{10{,}000}$ ________
8. $\frac{45{,}000}{5{,}000}$ ________

Use the formula $V = \frac{198}{d + 33}$ to solve these problems. (Need an example? See pages 62–63.)

What is the volume in cubic feet of a scientific balloon at each depth (d)?

9. $d = 33$ feet ________
10. $d = 0$ feet ________
11. $d = 165$ feet ________
12. $d = 66$ feet ________

Solve each equation. (Need an example? See pages 60–61.)

13. $\frac{a}{6} + 15 = 51$

____

14. $400 - 5b = 325$

____

15. $5c + 43.11 = 93.36$

____

16. $7d - 12.3 = 57.77$

____

17. $45 - 3h = 14$

____

18. $\frac{r}{3} + 12 = 18$

____

NAME ______

Daily Review 23

Write a pair of compatible numbers for each exercise. Then estimate each quotient. (Need an example? See pages 76–77.)

1. $330 \div 15$ ______
2. $1{,}018 \div 24$ ______
3. $4{,}673 \div 77$ ______
4. $89\overline{)16{,}000}$ ______
5. $47\overline{)52{,}894}$ ______
6. $108\overline{)82{,}314}$ ______

Divide using mental math. (Need an example? See pages 74–75.)

7. $3{,}500 \div 70$ ______
8. $18{,}000 \div 60$ ______
9. $540{,}000 \div 90$ ______
10. $41{,}000 \div 410$ ______
11. $\frac{810{,}000}{900}$ ______
12. $\frac{250{,}000}{500}$ ______

Solve each equation. (Need an example? See pages 60–61.)

13. $8x - 4.22 = 35.78$ ______
14. $\frac{y}{4} + 18 = 58$ ______
15. $12p + 54 = 534$ ______
16. $5r + 36.33 = 161.33$ ______
17. $8 + 3c = 29$ ______
18. $167 - 2e = 50$ ______

NAME

Daily Review 24

Divide using pencil and paper or calculator. Round decimal answers to the nearest hundredth.
(Need an example? See pages 78–79.)

1. 7,951 ÷ 81 ________ **2.** 1,905 ÷ 58 ________ **3.** 5,436 ÷ 15 ________

4. $53\overline{)5,492}$ **5.** $384\overline{)61,986}$ **6.** $50\overline{)87,314}$

Write a pair of compatible numbers for each exercise. Then estimate each quotient.
(Need an example? See pages 76–77.)

7. $89\overline{)44,347}$ ________________

8. $\frac{26,943}{296}$ ________________

9. $\frac{88,621}{284}$ ________________

Divide using mental math. (Need an example? See pages 74–75.)

10. 8,100 ÷ 90 ________ **11.** 450,000 ÷ 5,000 ________

12. 13,200 ÷ 110 ________ **13.** 900,000 ÷ 1,500 ________

14. 3,200 ÷ 20 ________ **15.** 42,000 ÷ 600 ________

16. 640 ÷ 80 ________ **17.** 72,000 ÷ 9,000 ________

NAME

Daily Review 25

Solve each problem. **Remember** to label the units in your answer. (Need an example? See pages 78–81.)

1. A group of 47 hikers separated into groups of 8. How many hikers were left over? ____________

2. How many 7-passenger boats are needed to transport 435 hikers across a river? ____________

3. A 4-mile trail is divided into 5 sections. How long is each section? ____________

4. Joe's team played 3 games last week. The team earned 21 points, 18 points, and 33 points in those games. Find the average number of points earned in a game. ____________

Write a pair of compatible numbers for each exercise. Then estimate each quotient. (Need an example? See pages 76–77.)

5. 490 ÷ 14 ____________________

6. 1,107 ÷ 23 ____________________

7. 11,360 ÷ 568 ____________________

Divide using mental math. (Need an example? See pages 74–75.)

8. 4,200 ÷ 60 __________ 9. 36,000 ÷ 900 __________

10. 540,000 ÷ 90 __________ 11. 700,000 ÷ 140 __________

NAME

Daily Review 26

Divide. Round decimal answers to the nearest hundredth. (Need an example? See pages 82–83.)

1. $46\overline{)281.51}$ **2.** $8\overline{)414.91}$ **3.** $7\overline{)229.32}$

4. $6\overline{)5.589}$ **5.** $67\overline{)279.29}$ **6.** $43\overline{)263.16}$

7. $3.249 \div 32$ ______ **8.** $4.003 \div 29$ ______ **9.** $389.6 \div 16$ ______

Solve each problem. **Remember** to label the units in your answer. (Need an example? See pages 80–81.)

10. There are 76 children who want to play baseball. There are 9 players on a team. How many children will be left over? ______

11. How many 21-passenger buses are needed to carry 456 children to the stadium? ______

12. The coach will individually instruct 6 players over the next 2 hours. He will spend the same amount of time with each player. How long will he spend with each? ______

NAME

Daily Review 27

Write each fraction as a decimal.
Remember to write zeros in the dividend.
(Need an example? See pages 84–85.)

1. $\frac{1}{4}$ ________ **2.** $\frac{5}{8}$ ________ **3.** $\frac{7}{25}$ ________

4. $\frac{3}{16}$ ________ **5.** $\frac{9}{200}$ ________ **6.** $\frac{7}{100}$ ________

7. $\frac{9}{40}$ ________ **8.** $\frac{15}{100}$ ________ **9.** $\frac{6}{500}$ ________

Divide. Round decimal answers to the nearest hundredth.
(Need an example? See pages 82–83.)

10. $59\overline{)345.97}$ **11.** $6\overline{)913.25}$ **12.** $42\overline{)1,104.6}$

13. $4.542 \div 5$ ________ **14.** $7.004 \div 20$ ________ **15.** $164.8 \div 16$ ________

Solve each problem. **Remember** to label the units in your answer. (Need an example? See pages 80–81.)

16. Ann and Arnold put the 74 rolls they baked into plastic bags. Six rolls went into each bag. How many rolls were left over? ________

17. How many 9-loaf trays are needed to transport 849 loaves of bread to the packaging area? ________

NAME ____________

Daily Review 28

Complete the table. **Remember** to move the decimal point to the right or to the left.
(Need an example? See pages 88–89.)

	k	$k \times 10$	$k \times 100$	$k \times 1{,}000$	$k \div 10$	$k \div 100$
1.	53.94					
2.	0.72					
3.	0.08					
4.	4.43					

Write each fraction as a decimal. Use a bar to indicate repeating digits. (Need an example? See pages 84–85.)

5. $\frac{2}{3}$ ____________ **6.** $\frac{5}{9}$ ____________ **7.** $\frac{3}{11}$ ____________

8. $\frac{5}{6}$ ____________ **9.** $\frac{4}{33}$ ____________ **10.** $\frac{26}{27}$ ____________

Divide. Round decimal answers to the nearest hundredth.
(Need an example? See pages 82–83.)

11. $47\overline{)238.19}$ **12.** $64\overline{)3.811}$ **13.** $83\overline{)9.003}$

14. $74\overline{)918.32}$ **15.** $46\overline{)118.3}$ **16.** $38\overline{)300.9}$

NAME

Daily Review 29

Divide. Round decimal answers to the nearest thousandth. **Remember** to place the decimal point in the quotient. (Need an example? See pages 90–91.)

1. $0.5\overline{)6.5}$ **2.** $7.5\overline{)0.675}$ **3.** $0.3\overline{)0.459}$

4. $39.6 \div 7.3$ ______ **5.** $4.408 \div 5.9$ ______ **6.** $68.32 \div 1.6$ ______

7. $3.94 \div 0.07$ ______ **8.** $20.7 \div 7.23$ ______ **9.** $391.5 \div 0.15$ ______

Complete the table. (Need an example? See pages 88–89.)

	k	$k \times 10$	$k \times 100$	$k \div 10$	$k \div 100$	$k \div 1{,}000$
10.	38.73					
11.	0.19					

Use a calculator and write each fraction as a decimal. Use a bar to indicate repeating decimals. (Need an example? See pages 84–85.)

12. $\frac{9}{60}$ ______ **13.** $\frac{14}{15}$ ______ **14.** $\frac{5}{12}$ ______

15. $\frac{1}{5}$ ______ **16.** $\frac{17}{33}$ ______ **17.** $\frac{23}{30}$ ______

18. $\frac{23}{33}$ ______ **19.** $\frac{1}{8}$ ______ **20.** $\frac{3}{4}$ ______

NAME

Daily Review 30

Did the person in each case make a wise and correct decision? (Need an example? See pages 92–93.)

1. Ed and Sarah were 80 miles away from home. They estimated that they had about 3 gallons of gas left. They know that they can get about 22 miles per gallon. They decided to fill their tank with gas. ______

2. Andy must type 8 pages. He usually takes 15 minutes to type one page. He has an hour and a half to finish the typing. He could have asked Jake to help with the typing but he decided against asking him. ______

Divide. Round decimal answers to the nearest thousandth. **Remember** to place the decimal point in the quotient. (Need an example? See pages 90–91.)

3. $0.4\overline{)8.9}$ 4. $3.7\overline{)0.283}$ 5. $6.4\overline{)1.472}$

6. $0.06 \div 4.69$ ______ 7. $50.20 \div 7.54$ ______ 8. $188.4 \div 0.25$ ______

Find each answer. (Need an example? See pages 88–89.)

9. $186.4 \times 1{,}000$ ______ 10. $87.42 \div 100$ ______ 11. $93.021 \div 10$ ______

NAME

Daily Review 31

Write an expression for each problem.
(Need an example? See pages 94–95.)

1. 4.5 times *a* ______
2. *b* divided by 6.2 ______
3. The product of *c* and 7.93 ______
4. 10.6 times *a* ______
5. 2.81 less than *d* ______
6. *p* divided by 8.9 ______
7. The sum of *r* and 25.62 ______
8. 5.03 less than *m* ______

Divide. Round answers to the nearest thousandth. **Remember** to place the decimal point in the quotient. (Need an example? See pages 90–91.)

9. $0.7 \div 0.004$ ______
10. $16.971 \div 0.8$ ______
11. $24.08 \div 0.14$ ______
12. $0.06 \div 5.93$ ______
13. $8.504 \div 6.7$ ______
14. $9.171 \div 0.09$ ______

Complete the table. (Need an example? See pages 88–89.)

	y	$y \times 10$	$y \times 100$	$y \times 1{,}000$	$y \div 10$	$y \div 100$
15.	72.18					
16.	0.64					

NAME ______________________

Daily Review 32

Solve each equation. Round decimal answers to the nearest thousandth.

Remember to line up the decimal points when adding or subtracting. (Need an example? See pages 96–97.)

1. $m + 2.81 = 5.40$ ______________________

2. $n - 12.9 = 3.76$ ______________________

3. $6.3p = 15.1$ ______________________

4. $0.84 = 11.4q$ ______________________

5. $\frac{r}{19.3} = 24.0$ ______________________

6. $5.47 = \frac{s}{1.7}$ ______________________

Write an expression for each problem. (Need an example? See pages 94–95.)

7. Quotient of 5.2 and *q* ______________________

8. Sum of *r* and 0.92 ______________________

9. 8.3 decreased by *s* ______________________

10. 3.78 more than *t* ______________________

11. Product of *c* and 0.37 ______________________

12. Sum of *v* and 43.62 ______________________

Did Fred make a wise and correct decision? (Need an example? See pages 92–93.)

13. Fred's car gets about 23 miles per gallon of gas. There are 5 gallons of gas in the tank and he is 110 miles from home. He decided that he had enough gas.

____

NAME

Daily Review 33

Write and solve an equation for each problem. **Remember** to check your work.
(Need an example? See pages 98–99.)

1. The Sanchez family is on its way to visit relatives 1,140 miles away. So far, they have covered 895 miles. How many more miles do they have to go?

______

2. It took the Sanchez family about 18 hours to cover 895 miles. What was their average speed?

______

Solve each equation. Round decimal answers to the nearest thousandth. **Remember** to line up decimal points when adding or subtracting.
(Need an example? See pages 96–97.)

3. $x + 1.8 = 4.25$

______

4. $\frac{y}{2.4} = 6$

______

5. $a - 3.05 = 4.7$

______

6. $12.4b = 4.96$

______

Write an expression for each problem. (Need an example? See pages 94–95.)

7. 8.9 times j

______

8. The sum of p and 1.42

______

9. 5.2 divided by q

______

10. k less 2.78

______

NAME

Daily Review 34

For Exercises 1–3, use the diagram at the right. (Need an example? See pages 114–117.)

1. Name three segments.

____

2. Name a pair of parallel lines.

____

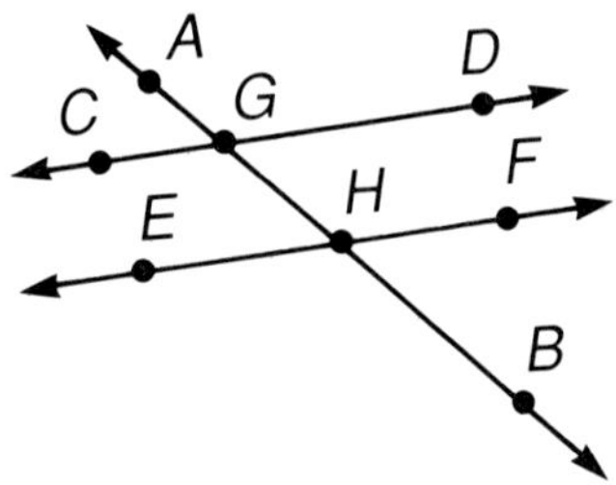

3. Name the intersection of $\overleftrightarrow{AB}$ and $\overleftrightarrow{EF}$.

____

Write and solve an equation for this problem. (Need an example? See pages 98–99.)

4. Joe determined that he spends about one third of his earnings on car care. If he spent \$120 on his car this month, how much did he earn?

____

Solve each equation. (Need an example? See pages 96–97.)

5. $125 = \frac{x}{2.5}$

____

6. $3.5s = 0.7$

____

7. $t - 11.8 = 32.94$

____

8. $y + 4.02 = 6.3$

____

NAME ____________

Daily Review 35

Use the diagram for Exercises 1–4. (Need an example? See pages 118–119.)

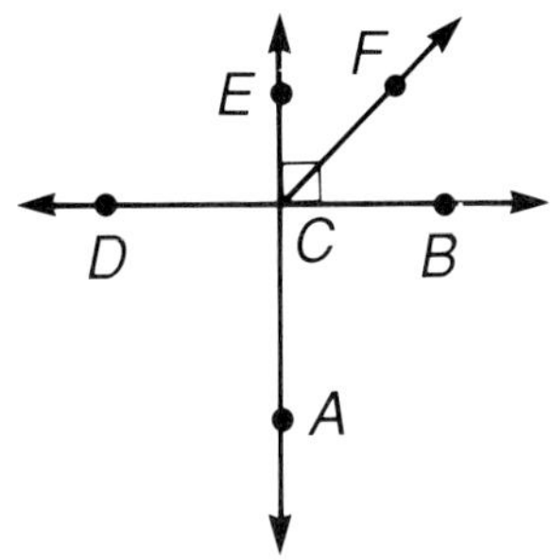

1. Name two acute angles.

____

2. Name two supplementary angles.

____

3. Name two complementary angles.

____

4. Name two right angles.

____

For Exercises 5 and 6, use the diagram at the right. (Need an example? See pages 114–117.)

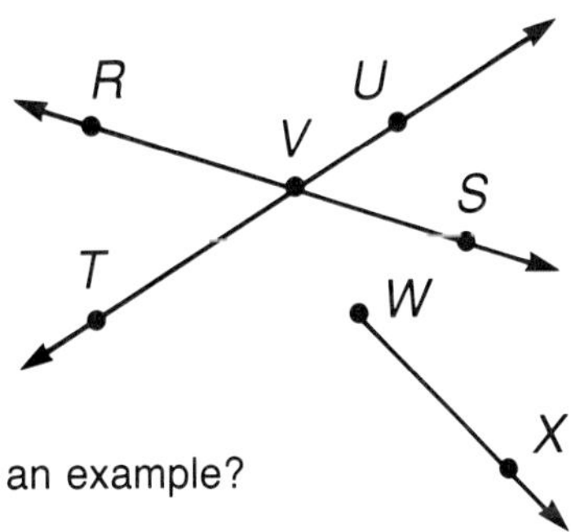

5. Name two rays.

____

6. Name two lines.

____

Write and solve an equation. (Need an example? See pages 98–99.)

7. Odetta received an increase in pay of 1.1 times her previous rate. She now earns $6.05 per hour. What was she earning before her raise?

____

  Use before pages 120–121.

NAME

Daily Review 36

Tell if the labels in Exercises 1–3 could be correct. Answer *yes* or *no*. (Need an example? See pages 120–121.)

1. 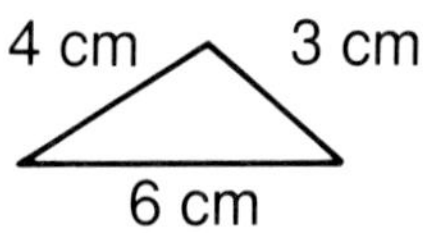

Scalene Triangle

______

2. 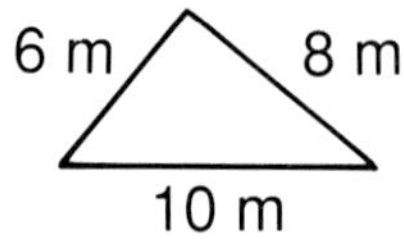

Isoceles Triangle

______

3. 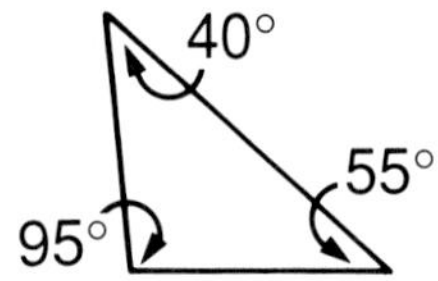

______

For Exercises 4 and 5, use the diagram at the right. (Need an example? See pages 118–119.)

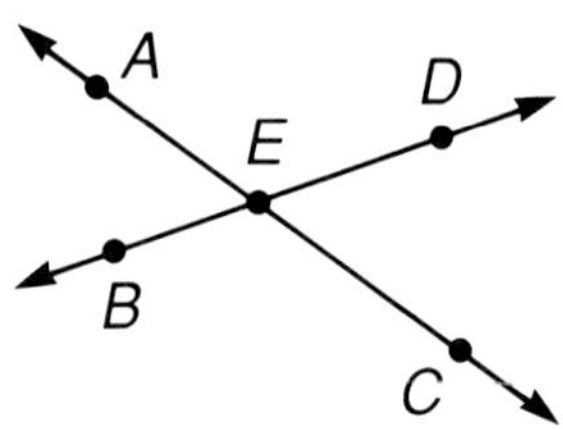

4. Name two obtuse angles.

____

5. Name two pairs of vertical angles.

____

For Exercises 6–8, use the diagram at the right. (Need an example? See pages 114–117.)

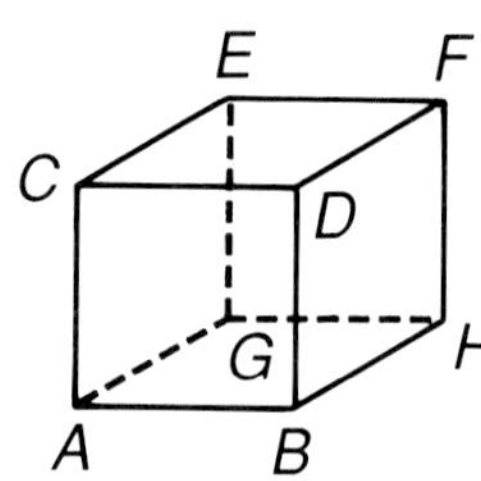

6. Name two parallel lines.

____

7. Name two skew lines.

____

8. Name the intersection of $\overleftrightarrow{CE}$ and $\overleftrightarrow{EF}$.

____

NAME

Daily Review 37

Draw the figure described. If the figure is impossible to draw, explain why. (Need an example? See pages 122–123.)

1. A parallelogram that is not a rhombus

2. A rectangle that is not a parallelogram

3. A quadrilateral that is not a parallelogram

Find the missing angle measure in each triangle. (Need an example? See pages 120–121.)

4.

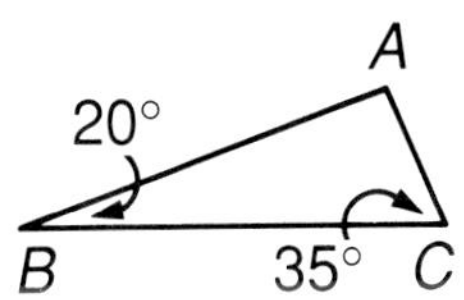

∠A ________

5.

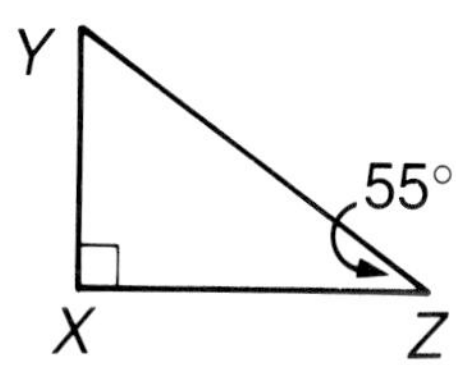

∠Y ________

6.

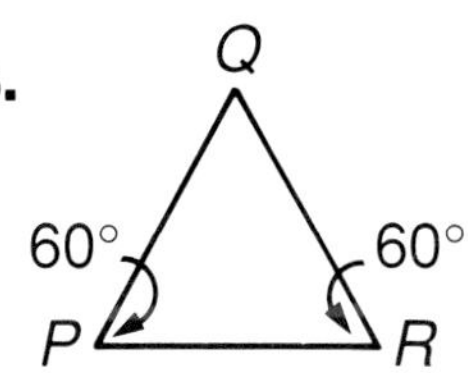

∠Q ________

For Exercises 7–9, use the diagram. (Need an example? See pages 118–119.)

7. Name two complementary angles.

____

8. Name two supplementary angles.

____

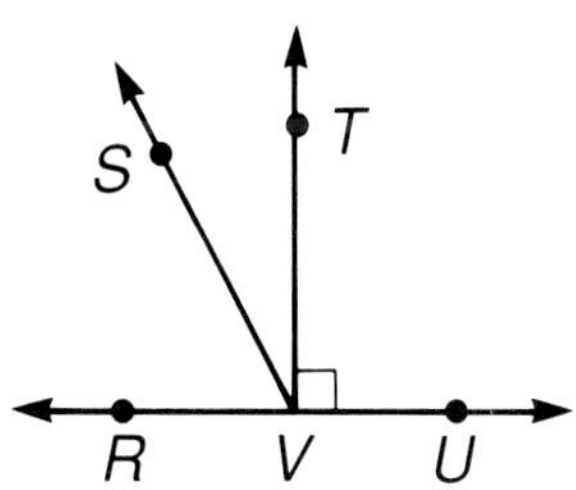

NAME ____________________

Daily Review 38

Make a table to solve each problem. (Need an example? See pages 124–125.)

1. Find the number of diagonals that can be drawn from all vertices in a hexagon. ____________

2. Find the sum of the angle measures in a polygon with 9 sides. ____________

Draw the figure described. If the figure is impossible to draw, explain why. (Need an example? See pages 122–123.)

3. A hexagon with all angles equal in measure but with sides that are not all equal in length

4. A square that is not a rhombus

Write which of these names can be used to describe each triangle below: *acute, right, obtuse, scalene, isosceles, equilateral.* (Need an example? See pages 120–121.)

5.

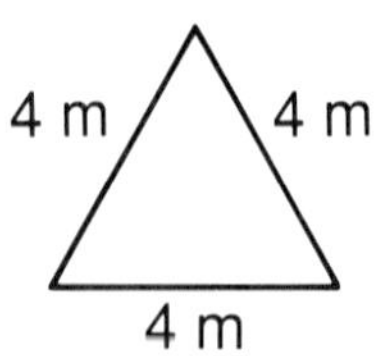

____

6.

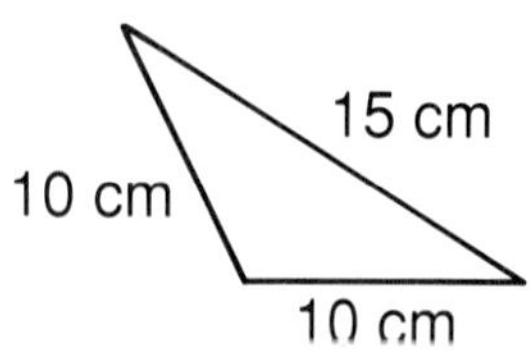

____

7.

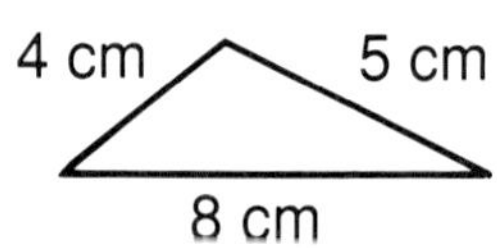

____

NAME

Daily Review 39

Use circle M for Exercises 1–6. (Need an example? See pages 126–127.)

1. Name the center. ______________

2. Name an arc.

____

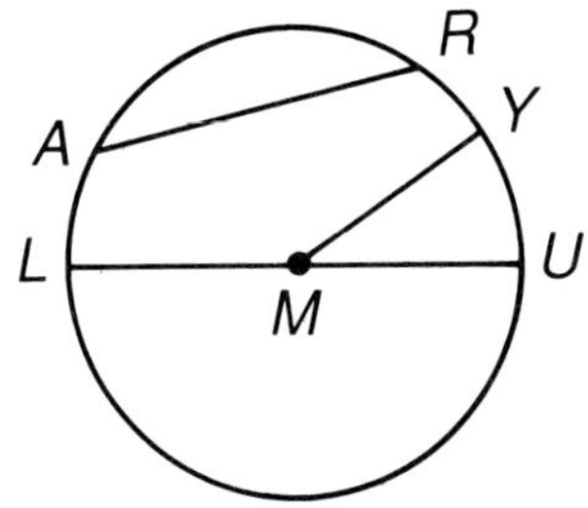

3. Name two central angles.

____

4. Name a diameter. ______________

5. Name two chords. ______________

Make a table to solve the problem. (Need an example? See pages 124–125.)

6. Find the sum of the angle measures in a polygon with 11 sides. ______________

Draw the figure described. If the figure is impossible to draw, explain why. (Need an example? See pages 122–123.)

7. A rhombus that is not a square

8. A pentagon that has 6 sides

NAME

Daily Review 40

On the blank grid to the right draw the result of each given translation. (Need an example? See pages 128–129.)

1. 2 units to the right

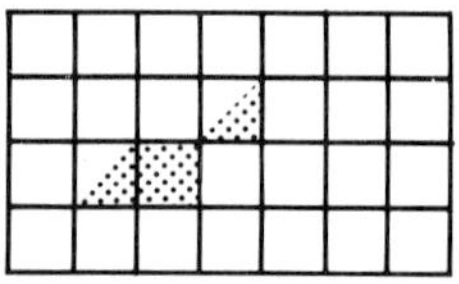

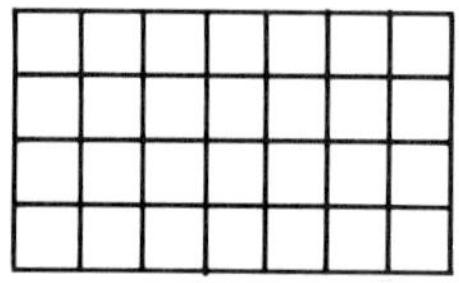

2. 2 units down; 3 units to the left

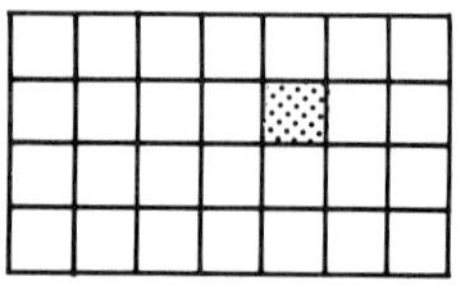

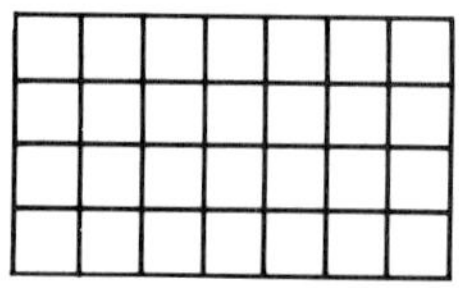

Use circle *P* for Exercises 3–5.
(Need an example? See pages 126–127.)

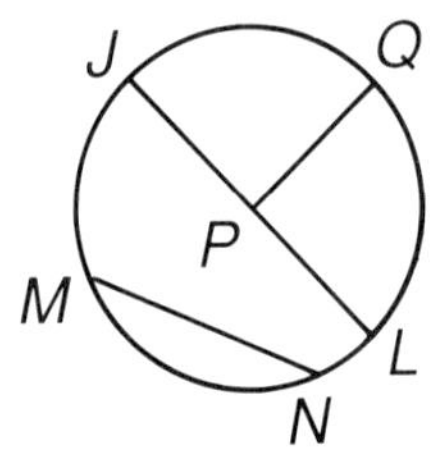

3. Name three radii.

____

4. Name two central angles.

____

5. Name two chords.

____

Make a table to solve the problem. (Need an example? See pages 124–125.)

6. Find the sum of the angle measures in a polygon with 10 sides. ________________

NAME ____________________

Daily Review 41

Draw all lines of symmetry on each letter.
(Need an example? See pages 130–133.)

1. 

2. 

3. 

On the blank grid to the right draw the result of the given translation. (Need an example? See pages 128–129.)

4. 3 units down

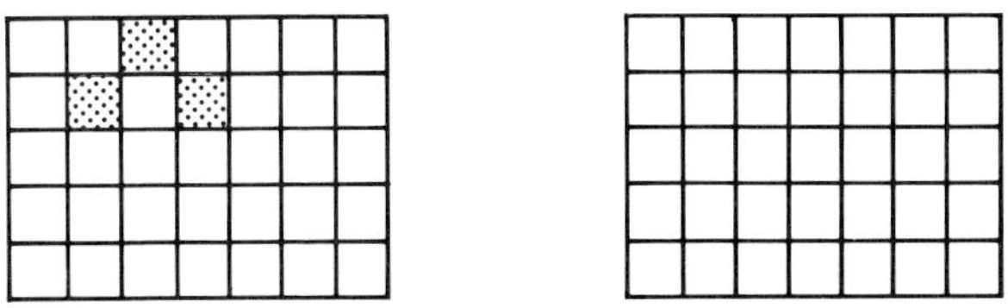

5. 2 units up; 3 units to the right

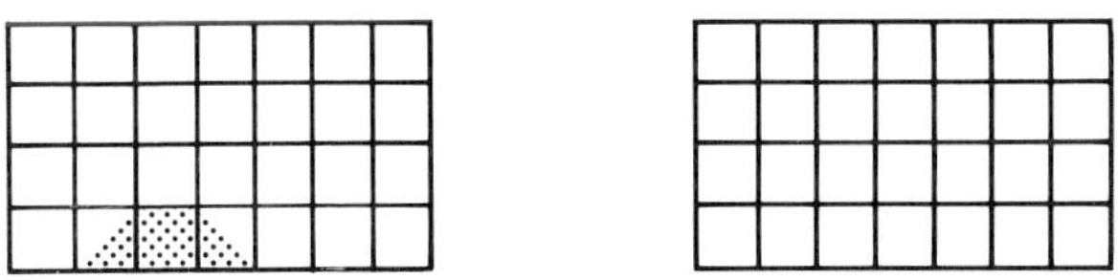

Use circle *G* for Exercises 6 and 7. (Need an example? See pages 126–127.)

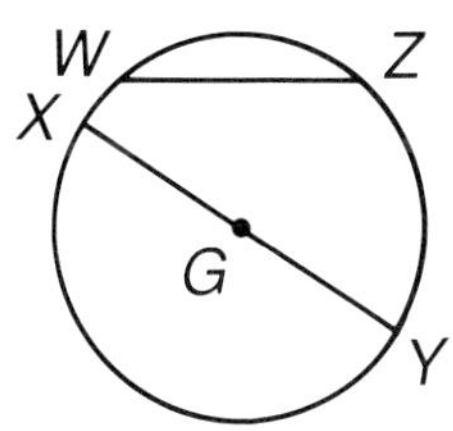

6. Name two arcs.

____

7. Name a chord that is not a diameter.

____

NAME

Daily Review 42

Use a compass and straightedge to do each construction. (Need an example? See pages 136–139.)

1. Construct a segment congruent to $\overline{EF}$.

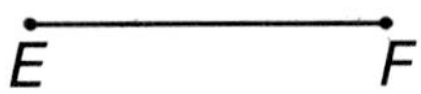

2. Bisect $\overline{EF}$.

Draw the reflection or rotation of each figure as described. (Need an example? See pages 130–133.)

3. Reflect over the line

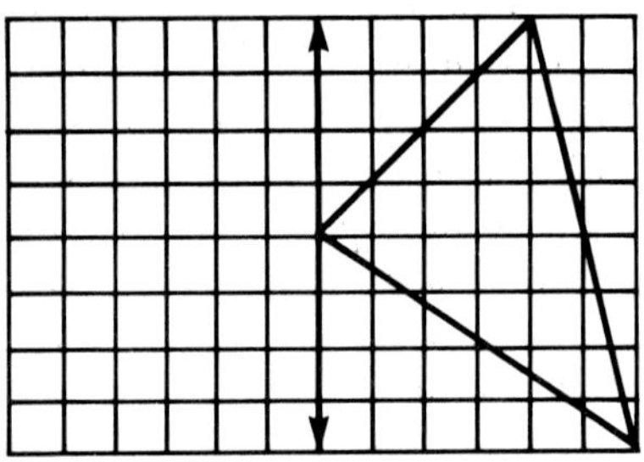

4. Rotate 90° about point *T*

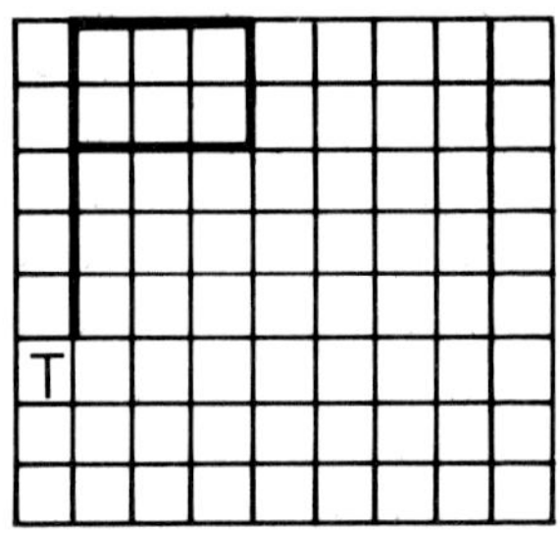

On the blank grid to the right draw the result of the given translation. (Need an example? See pages 128–129.)

5. 1 unit up; 3 units to the left

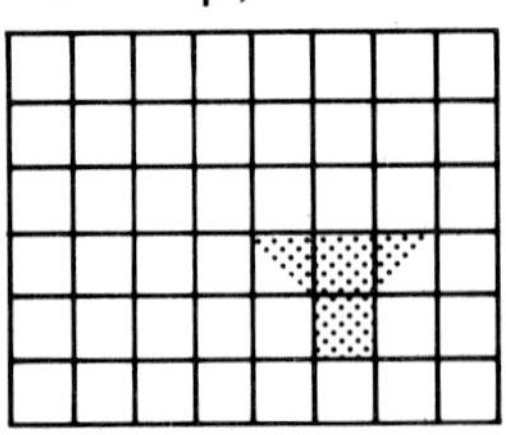

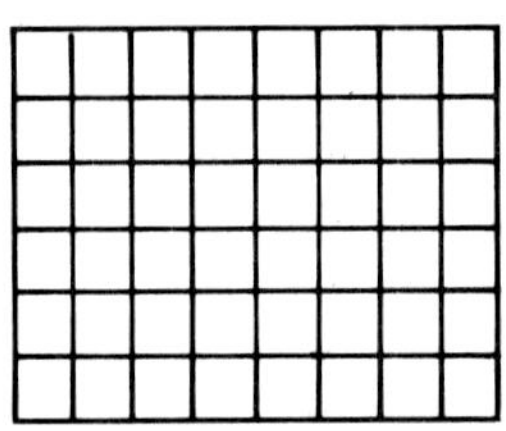

NAME

Daily Review 43

In Exercises 1 and 2, draw the figures described. Then do the constructions. (Need an example? See pages 140–141.)

1. Draw any $\overleftrightarrow{AB}$ and point G not on $\overleftrightarrow{AB}$. Construct the line through G parallel to $\overleftrightarrow{AB}$.

2. Draw any $\overleftrightarrow{PQ}$. Construct a line perpendicular to $\overleftrightarrow{PQ}$ at Q.

Use a compass and straightedge to do each construction. (Need an example? See pages 136–139.)

3. Construct an angle congruent to $\angle B$.

4. Bisect $\angle B$.

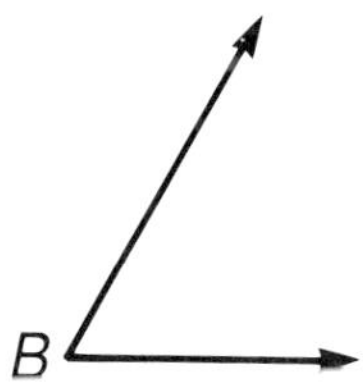

Draw all lines of symmetry on each figure. (Need an example? See pages 130–132.)

5.

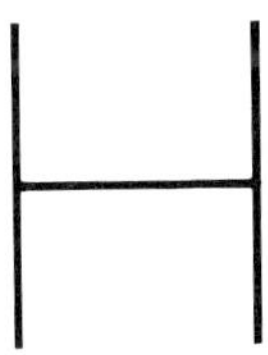

6.

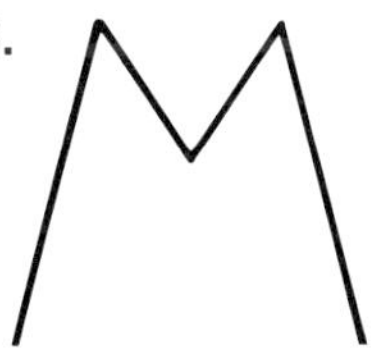

7.

NAME

Daily Review 44

Use triangle *ABC* for Exercises 1 and 2. Construct a congruent triangle using only the given parts. (Need an example? See pages 142–143.)

1. $\overline{AB}$, $\overline{BC}$, $\overline{AC}$

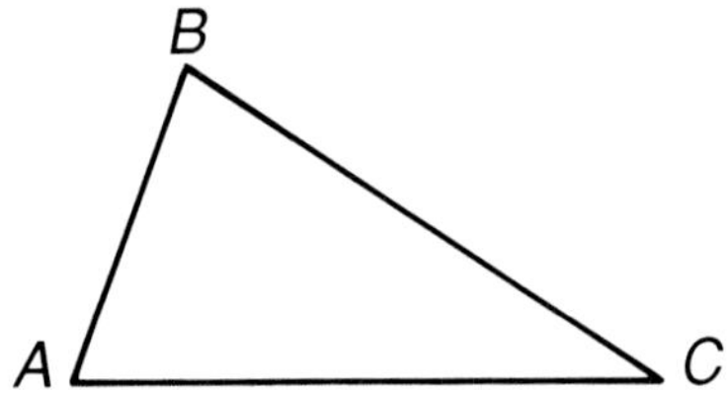

2. $\overline{AB}$, $\overline{AC}$, $\angle A$

In Exercises 3 and 4, draw the figures described. Then do the constructions. (Need an example? See pages 140–141.)

3. Draw any $\overleftrightarrow{DE}$ and a point *C* not on $\overleftrightarrow{DE}$. Construct the line from point *C* perpendicular to $\overleftrightarrow{DE}$.

4. Draw any $\overleftrightarrow{LM}$ and a point *R* not on $\overleftrightarrow{LM}$. Construct the line through *R* parallel to $\overleftrightarrow{LM}$.

Use a compass and straightedge to do each construction.
(Need an example? See pages 136–139.)

5. Bisect *JK*.

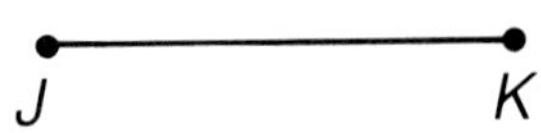

6. Bisect $\angle L$.

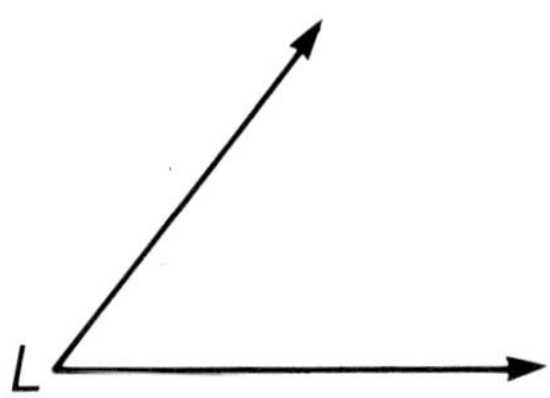

NAME

Daily Review 45

Solve each problem. The lengths of all sides must be whole numbers. (Need an example? See pages 144–145.)

1. The lengths of two sides of a triangle are 4 cm and 5 cm. What is the least possible length for the third side?

____

2. Two sides of a triangle have lengths 7 mm and 8 mm. List all possible lengths of the third side.

____

Use $\triangle PQR$ for Exercises 3 and 4. Construct the figure described. (Need an example? See pages 140–141.)

3. Square *DEFG* such that each side has the same length as $\overline{QR}$

Construct the figure described. (Need an example? See pages 142–143.)

4. $\triangle ABC$ such that each side has the same length as $\overline{PQ}$

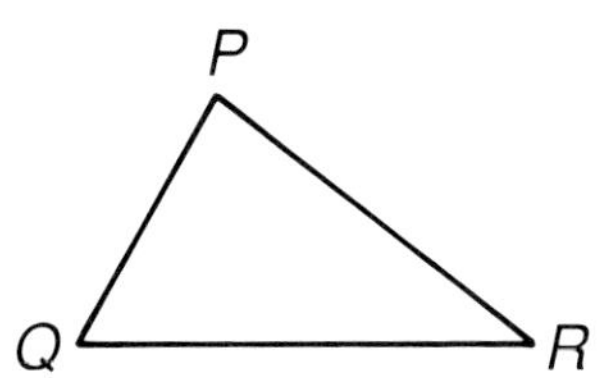

NAME

Daily Review 46

For each number, write if it is divisible by 2, 3, 5, 9, or 10. List all possibilities. Use mental math or pencil and paper. Write which method you used. (Need an example? See pages 156–157.)

	Divisible by:	Method Used:
1. 420	________	________
2. 5,062	________	________
3. 17,820	________	________
4. 56,381	________	________

Solve the problem. The lengths of all sides must be whole numbers. (Need an example? See pages 144–145.)

5. The lengths of two sides of a triangle are 11 mm and 29 mm. What is the least possible length for the third side? ________

Construct the triangle described. (Need an example? See pages 142–143.)

6. $\triangle JKL$ such that JK is twice MN, KL is twice NO, JL is twice MO.

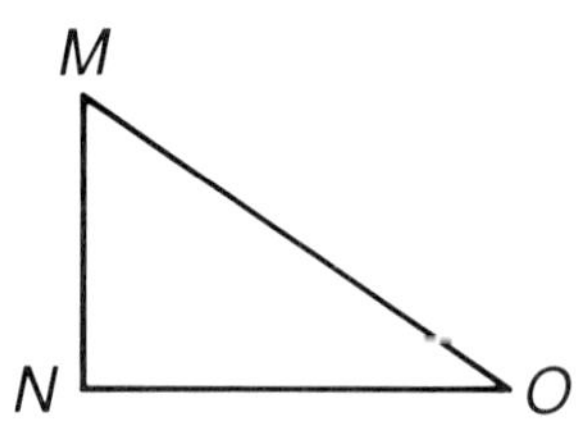

NAME

Daily Review 47

Write each power in standard form. Use mental math or pencil and paper. Write which method you used. (Need an example? See pages 158–159.)

1. 4^3 ____________ **2.** 2^4 ____________

3. 7^0 ____________ **4.** 3^3 ____________

5. 10^6 ____________ **6.** 5^3 ____________

For each number, write if it is divisible by 2, 3, 5, 9, or 10. List all possibilities. Use mental math or paper and pencil. Write which method you used.
(Need an example? See pages 156–157.)

	Divisible by:	Method Used:
7. 765	____________	____________
8. 10,008	____________	____________
9. 78,426	____________	____________

Solve the problems. The lengths of all sides must be whole numbers. (Need an example? See pages 144–145.)

10. Two sides of a triangle have lengths of 2 cm and 4 cm. List all possible lengths of the third side.

____

11. Two sides of a triangle have lengths of 11 cm. List all possible lengths of the third side.

____

NAME

Daily Review 48

Write *prime* or *composite.* (Need an example? See pages 160–163.)

1. 24 ____________ **2.** 19 ____________

3. 59 ____________ **4.** 69 ____________

5. 73 ____________ **6.** 102 ____________

7. 43 ____________ **8.** 91 ____________

9. 63 ____________ **10.** 441 ____________

Write each power in standard form. Use mental math or pencil and paper. Write which method you used. (Need an example? See pages 158–159.)

11. 9^2 ____________ **12.** 10^7 ____________

13. 16^1 ____________ **14.** 8^0 ____________

15. 6^4 ____________ **16.** 2^5 ____________

For each number, write if it is divisible by 2, 3, 5, 9, or 10. List all possibilities. Use mental math or paper and pencil. Write which method you used. (Need an example? See pages 156–157.)

	Divisible by:	Method Used:
17. 150	____________	____________
18. 35,070	____________	____________
19. 821	____________	____________
20. 6,001	____________	____________
21. 135	____________	____________

NAME

Daily Review 49

Make a factor tree for each number. Write the prime factorization for each number.
(Need an example? See pages 164–165.)

1. 48

____

2. 80

____

3. 28

____

4. 90

____

Write *prime* or *composite.*
(Need an example? See pages 160–163.)

5. 52 ______________________

6. 201 ______________________

7. 61 ______________________

8. 29 ______________________

Write each power in standard form. Use mental math or pencil and paper. Write which method you used. (Need an example? See pages 158–159.)

9. 2^7 ______________________

10. 10^9 ______________________

11. 4^5 ______________________

12. 7^3 ______________________

NAME ____________________

Daily Review 50

Find the GCF for each exercise. If you need help, use lists of factors. (Need an example? See pages 168–169.)

1. 10 and 12 ________
2. 36 and 45 ________
3. 28 and 49 ________
4. 42 and 54 ________
5. 50 and 125 ________
6. 39 and 48 ________

Write the prime factorization of each number using exponents. (Need an example? See pages 164–165.)

7. 63 ________
8. 72 ________
9. 250 ________
10. 600 ________
11. 26 ________
12. 54 ________

Write *prime* or *composite*.
(Need an example? See pages 160–163.)

13. 106 ________
14. 207 ________
15. 550 ________
16. 23 ________
17. 903 ________
18. 423 ________

Write each power in standard form. Use mental math or pencil and paper. Write which method you used. (Need an example? See pages 158–159.)

19. 3^4 ________
20. 2^9 ________
21. 8^3 ________
22. 6^3 ________
23. 5^4 ________
24. 4^5 ________

NAME

Daily Review 51

Find the LCM. Use a list of multiples. (Need an example? See pages 170–171.)

1. 2 and 5 ______________ **2.** 6 and 8 ______________

3. 9 and 10 ______________ **4.** 4 and 28 ______________

5. 15 and 20 ______________ **6.** 4 and 22 ______________

Find the GCF for each exercise. Use the prime factorization method. (Need an example? See pages 168–169.)

7. 50 and 15 ______________ **8.** 48 and 12 ______________

9. 36 and 54 ______________ **10.** 63 and 72 ______________

11. 18 and 23 ______________ **12.** 27 and 45 ______________

Write the prime factorization of each number using exponents. (Need an example? See pages 164–165.)

13. 126 ______________ **14.** 38 ______________

15. 105 ______________ **16.** 60 ______________

17. 150 ______________ **18.** 270 ______________

Write *prime* or *composite*. (Need an example? See pages 160–163.)

19. 49 ______________ **20.** 53 ______________

21. 79 ______________ **22.** 83 ______________

23. 99 ______________ **24.** 54 ______________

NAME ____________

Daily Review 52

Solve each problem. Make a table to find all the possible answers. (Need an example? See pages 172–173.)

1. Joe has $2.00 in quarters and dimes. What combinations of coins might he have?

2. There are 6 chairs in the room. Some have 3 legs and some have 4 legs. Gina counted 20 legs. How many of each kind of chair are there?

Find the LCM using prime factorization. (Need an example? See pages 170–171.)

3. 15 and 35 ____________
4. 20 and 25 ____________
5. 21 and 33 ____________
6. 9 and 11 ____________

Find the GCF for each exercise. Use the prime factorization method. (Need an example? See pages 168–169.)

7. 12 and 60 ____________
8. 42 and 63 ____________
9. 56 and 28 ____________
10. 22 and 46 ____________
11. 35 and 105 ____________
12. 44 and 121 ____________

Write the prime factorization of each number using exponents. (Need an example? See pages 164–165.)

13. 220 ____________
14. 525 ____________

NAME

Daily Review 53

Use Pascal's triangle to complete the table.
(Need an example? See pages 174–175.)

	Row	Sum of the First Four Numbers
1.	3	
2.	4	
3.	5	
4.	6	

Solve the problem. Make a table to find all the possible answers. (Need an example? See pages 172–173.)

5. Pencils cost 4¢ each. Pens cost 11¢ each. Elliot spent 53¢ on 8 items. How many pens and pencils did he buy? ________

Find the GCF of each pair of numbers. Then find the LCM of each pair of numbers. (Need an example? See pages 168–169 and 170–171.)

	GCF	LCM
6. 3 and 7		
7. 6 and 16		
8. 4 and 36		
9. 8 and 10		
10. 2 and 15		

NAME

Daily Review 54

Use logical reasoning to help you solve each problem.
(Need an example? See pages 176–177.)

1. Bottles of juice are packed 6 to a carton. Bottles of water are packed 8 to a carton. A shipment of 10 cartons contained 66 bottles. How many cartons of each type were included?

2. Pink erasers are sold 3 to a package. White erasers are sold 2 to a package. Jane bought 5 packages that included 12 erasers. How many packages of each type did she buy?

Complete the table, using Pascal's triangle. (Need an example? See pages 174–175.)

	Row	Number of Numbers
3.	3	
4.	4	
5.	5	
6.	6	

Solve the following problem. Make a table to find all possible answers. (Need an example? See pages 172–173.)

7. There are 25 students in Mrs. Moore's class. The students sit in groups of 3 or 4. What is the greatest possible number of groups in the class?

NAME ________________

Daily Review 55

Give each missing number. (Need an example? See pages 188–189.)

1. $\frac{1}{2} = \frac{\square}{16}$ **2.** $\frac{2}{3} = \frac{\square}{12}$ **3.** $\frac{4}{5} = \frac{20}{\square}$

4. $\frac{12}{16} = \frac{3}{\square}$ **5.** $\frac{24}{36} = \frac{2}{\square}$ **6.** $\frac{5}{6} = \frac{\square}{72}$

Use logical reasoning to help you solve each problem. (Need an example? See pages 176–177.)

7. Class 7A is divided into project teams of 3 or 4 students each. There are 7 teams and the total number of students is 27. How many teams of each size are there?

____

8. Macintosh apples were packed 6 to a bag. Delicious apples were packed 10 to a bag. Steve bought 9 bags of apples for his restaurant. He had 78 apples in all. How many bags of each type did he have?

____

Complete the table, using Pascal's triangle. (Need an example? See pages 174–175.)

	Row	Sum of the Numbers
9.	4	
10.	5	
11.	6	
12.	7	

NAME

Daily Review 56

Write each fraction as a whole number or as a mixed number. **Remember** to write the fraction in the mixed number in lowest terms. (Need an example? See pages 190–191.)

1. $\frac{5}{2}$ ________ **2.** $\frac{64}{8}$ ________ **3.** $\frac{23}{7}$ ________

4. $\frac{8}{3}$ ________ **5.** $\frac{16}{12}$ ________ **6.** $\frac{54}{9}$ ________

7. $\frac{12}{3}$ ________ **8.** $\frac{42}{8}$ ________ **9.** $\frac{32}{7}$ ________

Is each fraction in lowest terms? Write *yes* or *no.* (Need an example? See pages 188–189.)

10. $\frac{12}{15}$ ________ **11.** $\frac{6}{9}$ ________ **12.** $\frac{3}{13}$ ________

13. $\frac{16}{21}$ ________ **14.** $\frac{18}{39}$ ________ **15.** $\frac{4}{34}$ ________

Use logical reasoning to help you solve the problems. (Need an example? See pages 176–177.)

16. Bread rolls are sold in packages of 6 and packages of 8. The team bought 12 packages for their end-of-season picnic. This gave them a total of 92 rolls. How many of each kind of package did they buy?

____

17. Basketball teams have 5 members and baseball teams have 9 members. At a camp 76 students were assigned to either a basketball team or a baseball team. There were 12 complete teams. How many teams of each type were there?

____

NAME ______________________

Compare the numbers. Use $<$, $>$, or $=$.

(Need an example? See pages 192–193.)

1. $\frac{4}{9} \bigcirc \frac{7}{9}$ **2.** $\frac{5}{6} \bigcirc \frac{10}{12}$ **3.** $\frac{4}{5} \bigcirc \frac{7}{10}$

4. $\frac{3}{4} \bigcirc \frac{7}{10}$ **5.** $\frac{4}{5} \bigcirc \frac{3}{5}$ **6.** $\frac{2}{11} \bigcirc \frac{8}{9}$

7. $2\frac{7}{20} \bigcirc 3\frac{1}{10}$ **8.** $6\frac{2}{3} \bigcirc 6\frac{3}{4}$ **9.** $1\frac{1}{6} \bigcirc 1\frac{1}{5}$

Write each number as a fraction.

(Need an example? See pages 190–191.)

10. $2\frac{2}{3}$ ________ **11.** $7\frac{1}{4}$ ________ **12.** $3\frac{2}{5}$ ________

13. $4\frac{1}{2}$ ________ **14.** $10\frac{1}{6}$ ________ **15.** $8\frac{4}{9}$ ________

16. $5\frac{1}{4}$ ________ **17.** $3\frac{1}{8}$ ________ **18.** $10\frac{3}{10}$ ________

Give each missing number.

(Need an example? See pages 188–189.)

19. $\frac{3}{5} = \frac{\square}{30}$ **20.** $\frac{2}{7} = \frac{\square}{21}$ **21.** $\frac{5}{6} = \frac{35}{\square}$

22. $\frac{10}{12} = \frac{\square}{6}$ **23.** $\frac{8}{36} = \frac{2}{\square}$ **24.** $\frac{40}{62} = \frac{20}{\square}$

25. $\frac{18}{24} = \frac{\square}{4}$ **26.** $\frac{20}{30} = \frac{2}{\square}$ **27.** $\frac{35}{40} = \frac{\square}{8}$

 Use before pages 194–195.

NAME

Daily Review 58

Choose the *best* estimate by comparing each fraction to $\frac{1}{2}$. Is the sum: (a) greater than 1, (b) less than 1, or (c) greater than the greater fraction in the exercise? (Need an example? See pages 194–195.)

1. $\frac{2}{7} + \frac{2}{5}$ ______ **2.** $\frac{5}{6} + \frac{3}{4}$ ______ **3.** $\frac{4}{5} + \frac{1}{8}$ ______

4. $\frac{3}{10} + \frac{1}{4}$ ______ **5.** $\frac{1}{6} + \frac{1}{4}$ ______ **6.** $\frac{6}{7} + \frac{1}{10}$ ______

7. $\frac{3}{5} + \frac{7}{12}$ ______ **8.** $\frac{3}{4} + \frac{4}{5}$ ______ **9.** $\frac{2}{3} + \frac{2}{5}$ ______

10. $\frac{3}{4} + \frac{2}{9}$ ______ **11.** $\frac{4}{5} + \frac{5}{8}$ ______ **12.** $\frac{3}{7} + \frac{7}{10}$ ______

List the numbers from least to greatest. (Need an example? See pages 192–193.)

13. $\frac{2}{3}, \frac{5}{6}, \frac{3}{4}$ ______ **14.** $\frac{4}{9}, \frac{1}{3}, \frac{1}{6}$ ______

15. $4\frac{1}{2}, 4\frac{2}{3}, 3\frac{5}{6}$ ______ **16.** $8\frac{3}{5}, 9, 7\frac{7}{10}$ ______

Write each fraction as a whole number or as a mixed number. **Remember** to write the fraction in the mixed number in lowest terms. (Need an example? See pages 190–191.)

17. $\frac{4}{2}$ ______ **18.** $\frac{9}{4}$ ______ **19.** $\frac{13}{5}$ ______

20. $\frac{10}{8}$ ______ **21.** $\frac{24}{6}$ ______ **22.** $\frac{45}{10}$ ______

23. $\frac{11}{3}$ ______ **24.** $\frac{21}{6}$ ______ **25.** $\frac{64}{5}$ ______

NAME

Daily Review 59

Add. (Need an example? See pages 196–199.)

1. $4\frac{2}{3} + 4\frac{2}{3}$

2. $3\frac{1}{8} + 7\frac{5}{8}$

3. $8\frac{1}{6} + 2\frac{1}{6}$

4. $\frac{1}{3} + \frac{5}{6}$

5. $\frac{3}{4} + \frac{1}{2}$

6. $\frac{3}{8} + \frac{3}{4}$

7. $5\frac{2}{5} + 1\frac{4}{5}$

8. $\frac{2}{3} + \frac{1}{4}$

9. $3\frac{1}{5} + \frac{2}{3}$

Estimate. Explain how you estimated. (Need an example? See pages 194–195.)

10. $\frac{1}{3} + \frac{2}{5}$ ____________

11. $\frac{3}{4} + \frac{7}{8}$ ____________

12. $9\frac{1}{4} - 2\frac{3}{5}$ ____________

13. $16\frac{2}{3} - 4\frac{1}{2}$ ____________

Compare the numbers. Use $<$, $>$, or $=$. (Need an example? See pages 192–193.)

14. $\frac{3}{5}$ ◯ $\frac{1}{2}$

15. $\frac{7}{8}$ ◯ $\frac{3}{4}$

16. $\frac{2}{3}$ ◯ $\frac{4}{9}$

17. $3\frac{1}{3}$ ◯ $3\frac{5}{6}$

18. $2\frac{2}{5}$ ◯ $1\frac{4}{5}$

19. $5\frac{8}{9}$ ◯ $5\frac{1}{4}$

NAME

Daily Review 60

Subtract.

(Need an example? See pages 202–203.)

1. $\frac{7}{8} - \frac{3}{4}$

2. $\frac{2}{3} - \frac{1}{6}$

3. $\frac{7}{9} - \frac{1}{3}$

4. $\frac{5}{8} - \frac{1}{6}$

5. $6\frac{5}{6} - 1\frac{2}{3}$

6. $4\frac{7}{10} - 2\frac{2}{5}$

In Exercises 7–9, tell whether you would use mental math or paper and pencil. Then find each sum. **Remember** to look for fractions that have a sum of 1. (Need an example? See pages 196–199.)

7. $\frac{1}{6} + \frac{5}{6}$

8. $\frac{7}{10} + \frac{3}{10}$

9. $3\frac{7}{12} + 4\frac{5}{12}$

Choose the *best* estimate by comparing each fraction to $\frac{1}{2}$. Is the sum: (a) greater than 1, (b) less than 1, or (c) greater than the greater fraction in the exercise? (Need an example? See pages 194–195.)

10. $\frac{8}{9} + \frac{3}{5}$ ______

11. $\frac{3}{10} + \frac{6}{7}$ ______

12. $\frac{1}{3} + \frac{2}{7}$ ______

NAME

Daily Review 61

Subtract. **Remember** to write answers in lowest terms. (Need an example? See pages 202 and 204.)

1. $6\frac{1}{3} - 2\frac{2}{3}$

2. $4\frac{3}{8} - 3\frac{1}{2}$

3. $7 - 2\frac{1}{4}$

4. $9 - 3\frac{7}{8}$

5. $11\frac{1}{5} - 9\frac{3}{10}$

6. $4 - 1\frac{2}{3}$

7. $\frac{7}{10} - \frac{3}{5}$

8. $\frac{3}{4} - \frac{1}{2}$

9. $\frac{5}{6} - \frac{3}{10}$

Add. (Need an example? See pages 196–199.)

10. $4\frac{1}{5} + 2\frac{3}{5}$

11. $2\frac{3}{8} + 7\frac{7}{8}$

12. $1\frac{5}{6} + 3\frac{1}{6}$

13. $3\frac{1}{8} + 2\frac{6}{8} =$ ___

14. $4\frac{1}{3} + 3\frac{2}{3} =$ ___

15. $1\frac{15}{16} + \frac{3}{16} =$ ___

16. $\frac{1}{3} + \frac{1}{6}$

17. $\frac{2}{9} + \frac{1}{3}$

18. $\frac{1}{4} + \frac{2}{5}$

NAME

Daily Review 62

Write an equation for each problem. Then solve the equation. (Need an example? See pages 206–207.)

1. Mrs. Richardson cut $2\frac{1}{2}$ ft from a board so that it could be used as a shelf in her closet. The length of the shelf is $4\frac{1}{4}$ ft. How long was the board before she cut it?

____

2. The total length of a quilt coverlet is 54 in. There are $6\frac{1}{2}$ in. of lace trim at the bottom. How long is the quilted fabric?

____

Subtract. **Remember** to write answers in lowest terms.
(Need an example? See pages 202 and 204.)

3. $4\frac{1}{5} - 2\frac{7}{10}$

4. $6\frac{3}{8} - 3\frac{1}{2}$

5. $11 - 5\frac{1}{4}$

6. $6\frac{5}{8} - 2\frac{1}{4}$

7. $20 - 5\frac{2}{3}$

8. $14\frac{1}{2} - 6\frac{2}{3}$

9. $\frac{3}{4} - \frac{1}{6} =$ ___

10. $\frac{5}{8} - \frac{1}{4} =$ ___

11. $4\frac{1}{5} - 1\frac{2}{5} =$ ___

12. $\frac{7}{8} - \frac{1}{3} =$ ___

13. $\frac{9}{10} - \frac{1}{2} =$ ___

14. $5\frac{1}{4} - 3\frac{3}{4} =$ ___

NAME

Daily Review 63

Multiply. Use the shortcut method when possible. (Need an example? See pages 208–211.)

1. $\frac{4}{5} \times \frac{1}{4} =$ ____ **2.** $\frac{3}{4} \times \frac{2}{3} =$ ____ **3.** $\frac{1}{6} \times \frac{2}{5} =$ ____

4. $\frac{11}{12} \times 0 =$ ____ **5.** $7 \times \frac{2}{3} =$ ____ **6.** $\frac{1}{3} \times 1\frac{5}{6} =$ ____

Write an equation for each problem. Then solve the equation. (Need an example? See pages 206–207.)

7. A class banner is 6 ft long. About $2\frac{1}{3}$ ft of it is blue silk. The rest is white silk. How much white silk is used for the banner?

____

8. A dresser that is $4\frac{1}{4}$ ft high fits into a closet. If the dresser leaves a space of $2\frac{1}{3}$ ft above it, what is the height of the closet?

____

Subtract. **Remember** to write answers in lowest terms.
(Need an example? See pages 204–205.)

9. $7\frac{1}{3} - 3\frac{5}{6}$

10. $8 - 4\frac{5}{8}$

11. $12\frac{1}{2} - 5\frac{2}{3}$

12. $5\frac{1}{8} - 2\frac{7}{8}$

13. $10 - 3\frac{1}{10}$

14. $7\frac{1}{3} - 1\frac{3}{4}$

NAME

Daily Review 64

Solve each problem. (Need an example? See pages 212–213.)

1. Kyle is training for a track meet. If he runs 10 more laps today, he will be halfway through the 24 he must do. How many laps has he done so far? __________

2. Joan spent $\frac{1}{2}$ of the time between dinner and bedtime on homework. She spent $\frac{1}{4}$ of the remaining time reading. Then she spent $\frac{2}{3}$ of the rest of the time working on an art project. That left 30 minutes to get ready for bed. How much time did she have to start with? __________

Multiply. Tell whether you use paper and pencil or mental math. (Need an example? See pages 208–211.)

3. $\frac{7}{10} \times \frac{2}{3} =$ __________

4. $1 \times \frac{23}{24} =$ __________

5. $\frac{2}{3} \times \frac{1}{8} \times \frac{3}{4} =$ __________

6. $\frac{6}{7} \times \frac{4}{9} \times \frac{1}{3} =$ __________

Write an equation for the problem. Then solve the equation. (Need an example? See pages 206–207.)

7. Rose added a row of bricks to an existing row that measured $4\frac{1}{2}$ ft. The new row measured 10 ft. How long was Rose's row?

__________ __________

NAME

Daily Review 65

Divide. **Remember** to divide by a number, multiply by its reciprocal. (Need an example? See pages 214–217.)

1. $6 \div \frac{2}{3} =$ ____ **2.** $5 \div \frac{3}{10} =$ ____ **3.** $\frac{1}{4} \div 1\frac{1}{2} =$ ____

4. $\dfrac{\frac{2}{5}}{\frac{1}{10}}$ ____ **5.** $\dfrac{\frac{3}{4}}{\frac{1}{8}}$ ____ **6.** $\dfrac{2}{1\frac{1}{8}}$ ____

Solve each problem. (Need an example? See pages 212–213.)

7. Marla sold $\frac{1}{2}$ of the tickets to her school's fair in one week. She sold $\frac{3}{4}$ of the remaining tickets the following week. There are 100 tickets left to sell. How many did she have to sell to begin with? ____

8. Ken has to pack 15 boxes. When he packs 3 more boxes, he will have completed $\frac{2}{3}$ of the job. How many boxes has he packed so far? ____

Multiply. Find the range to see whether your answers are reasonable. (Need an example? See pages 208–211.)

9. $3\frac{1}{2} \times 8 =$ ____ **10.** $6 \times 2\frac{1}{3} =$ ____ **11.** $2\frac{1}{4} \times 4\frac{1}{3} =$ ____

Range: ____ ____ ____

12. $5\frac{1}{2} \times 2 =$ ____ **13.** $4\frac{2}{3} \times 6\frac{1}{2} =$ ____ **14.** $8 \times 5\frac{1}{4} =$ ____

Range: ____ ____ ____

NAME

Daily Review 66

Solve each equation. (Need an example? See pages 218–219.)

1. $\frac{1}{4}a = 11$ ______

2. $6 = 1\frac{1}{2}b$ ______

3. $\frac{c}{8} = 2\frac{1}{8}$ ______

4. $6 = \frac{3}{4}d$ ______

5. $3e = \frac{4}{5}$ ______

6. $\frac{1}{2}f = 2\frac{7}{10}$ ______

Divide. **Remember** to divide by a number, multiply by its reciprocal. (Need an example? See pages 214–217.)

7. $6 \div \frac{2}{3} =$ ______

8. $4 \div \frac{3}{8} =$ ______

9. $3\frac{3}{10} \div 11 =$ ______

10. $2\frac{2}{3} \div 1\frac{1}{7} =$ ______

11. $2\frac{1}{4} \div 3\frac{3}{5} =$ ______

12. $3\frac{4}{7} \div 5 =$ ______

13. $2\frac{1}{2} \div 5 =$ ______

14. $10 \div 2\frac{1}{2} =$ ______

Solve the problem. (Need an example? See pages 212–213.)

15. Louise promised her dad she would plant 27 tomato plants. When she has planted 4 times as many as she has now done, she will have only 3 more to plant. How many has she planted already? ______

NAME

Daily Review 67

Explain how the terms are related. Then write the next three terms. (Need an example? See pages 220–221.)

1. $1 \quad \frac{1}{10} \quad \frac{1}{100}$ ______________

______ ______ ______

2. $\frac{1}{3} \quad 2 \quad 3\frac{2}{3}$ ______________

______ ______ ______

3. $2 \quad 1\frac{3}{4} \quad 1\frac{1}{2}$ ______________

______ ______ ______

4. $\frac{1}{2} \quad \frac{1}{8} \quad \frac{1}{32}$ ______________

______ ______ ______

5. Write the next two entries in the table for the decrease in radioactive tritium.

Years	0	$12\frac{1}{2}$	25		
Amount Left	1	$\frac{1}{2}$	$\frac{1}{4}$		

Solve each equation. (Need an example? See pages 218–219.)

6. $\frac{1}{4}x = 12$ ______

7. $\frac{2}{3}x = 18$ ______

8. $4x = \frac{3}{4}$ ______

9. $2\frac{1}{5}x = 33$ ______

10. $5x = 1\frac{2}{3}$ ______

11. $1\frac{1}{2}x = 2\frac{1}{4}$ ______

Divide. (Need an example? See pages 214–217.)

12. $\dfrac{2\frac{4}{5}}{7}$ ______

13. $\dfrac{10}{1\frac{2}{5}}$ ______

14. $\dfrac{\frac{3}{4}}{18}$ ______

15. $\dfrac{9}{2\frac{1}{6}}$ ______

NAME

Daily Review 68

Choose the best measure. (Need an example? See pages 232–235.)

1. The length of a pencil
   a. 7 in. **b.** 7 ft **c.** 7 yd ____
2. The width of a car
   a. 5 in. **b.** 5 ft **c.** 5 yd ____
3. The area of a blanket
   a. 24 sq in. **b.** 24 sq ft **c.** 24 sq yd ____
4. The volume of a mug
   a. 2 cu in. **b.** 10 cu in. **c.** 60 cu in. ____

Explain how the terms are related. Then write the next three terms. (Need an example? See pages 220–221.)

5. $1\frac{1}{10}$ $2\frac{1}{5}$ $3\frac{3}{10}$ ____

   ____ ____ ____

6. 2 $\frac{2}{3}$ $\frac{2}{9}$ ____

   ____ ____ ____

7. $\frac{3}{5}$ $\frac{9}{50}$ $\frac{27}{500}$ ____

   ____ ____ ____

8. 4 $3\frac{1}{4}$ $2\frac{1}{2}$ ____

   ____ ____ ____

Solve each equation. (Need an example? See pages 218–219.)

9. $2\frac{1}{3}x = 7$ ____
10. $3x = 4\frac{1}{4}$ ____
11. $1\frac{2}{9}x = 2\frac{1}{5}$ ____
12. $2\frac{1}{2}x = 15$ ____
13. $4\frac{1}{5}x = 21$ ____
14. $3\frac{1}{3}x = 5\frac{1}{4}$ ____

NAME ____________________

Daily Review 69

Find each missing number. Use paper and pencil or a calculator. (Need an example? See pages 236–237.)

1. 3 qt = ___ pt
2. 56 qt = ______ gal
3. 5 pt = ____ qt ____ pt
4. 6 T = ______ lb
5. 80 oz = ______ lb
6. 7,800 lb = ___ T _____ lb
7. 9 c = ______ fl oz
8. 6 lb 4 oz = _____ oz

Find each missing number. Use the tables on pages 232–233. (Need an example? See pages 232–235.)

9. 4 ft = ___ in.
10. 60 in. = ___ ft
11. 32 ft = ___ yd ___ ft
12. 3 sq yd = ___ sq ft
13. 45 sq ft = ___ sq yd
14. 2 cu ft = ____ cu in.

Explain how the terms are related. Then write the next three terms. (Need an example? See pages 220–221.)

15. 1 $\frac{1}{4}$ $\frac{1}{16}$ ______________

    ______ ______ ______

16. $5\frac{1}{3}$ $4\frac{2}{3}$ 4 ______________

    ______ ______ ______

17. 2 $3\frac{2}{3}$ $5\frac{1}{3}$ ______________

    ______ ______ ______

18. $\frac{2}{5}$ $\frac{4}{20}$ $\frac{8}{80}$ ______________

    ______ ______ ______

19. 9 $8\frac{1}{5}$ $7\frac{2}{5}$ ______________

    ______ ______ ______

20. 10 $\frac{20}{3}$ $\frac{40}{9}$ ______________

    ______ ______ ______

NAME ____________________

Daily Review 70

Use information from the Fact Sheet. If there is not enough information given to solve the problems, write *too little information.*

(Need an example? See pages 238–239.)

FACT SHEET
1 cu in. of water weighs 0.554 oz.
1 cu ft of water weighs 62.4 lb

1. A rectangular vase measures 2 in. by 3 in. by 10 in. How much does the water that fills the vase weigh?

____

2. How much does 40 cu in. of maple syrup weigh?

____

Find each missing number. Use paper and pencil or a calculator. (Need an example? See pages 236–237.)

3. 7 qt = ______ pt

4. 10 qt = ______ gal ______ qt

5. 320 oz = ______ lb

6. 45 c = ______ pt ______ c

Find the area. (Need an example? See pages 232–235.)

7.

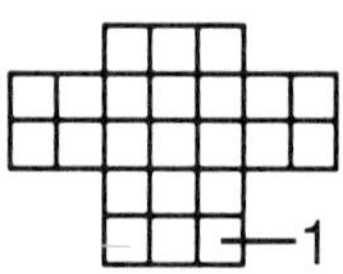

—1 sq in.

____

8.

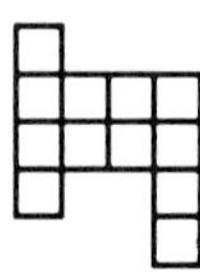

—1 sq in.

____

NAME

Daily Review 71

Complete each unit of measure with either *gram, liter,* or *meter.* (Need an example? See pages 240–241.)

1. Capacity of a cup: milli ____________

2. Weight of a man: kilo ____________

3. Distance between cities: kilo ____________

4. Length of an ant: milli ____________

Use the information from the Fact Sheet to solve the problems. If there is not enough information given, write *too little information.* (Need an example? See pages 238–239.)

FACT SHEET
1 tablespoon = 3 teaspoons
1 gal = 231 cu in.

5. A jar holds 23 teaspoons of baking powder. How many tablespoons does it hold?

____

6. A bottle has a volume of 462 cu in. How many quarts can it hold?

____

Find each missing number. Use paper and pencil or a calculator. (Need an example? See pages 236–237.)

7. 5 c = ______ fl oz

8. 33 gal = ______ qt

9. 7 pt = ______ qt ______ pt

10. 9 T = ____________ lb

11. 24 lb = ______ oz

12. 32 oz = ______ lb

NAME ______

Daily Review 72

Find each missing number. (Need an example? See pages 242–245.)

1. 65 dm = ______ m

2. 0.9 m = ______ mm

3. 4.7 km = ______ m

4. 315 mm = ______ cm

5. 52 m = ______ cm

6. 25 cm = ______ m

Use the table on page 240 to find each missing number. (Need an example? See pages 240–241.)

7. 1 km = ______ m

8. 1 mm = ______ cm

9. 1 g = ______ kg

10. 1 dm = ______ cm

11. 1 L = ______ kL

12. 1 mL = ______ L

Use information from the Fact Sheet to solve the problems. If there is not enough information given to solve the problems, write *too little information.* (Need an example? See pages 238–239.)

13. How far is Houston from Philadelphia?

______

14. How much farther away from Los Angeles is New York than Philadelphia?

______

FACT SHEET

Road Distance

Los Angeles to New York: 2,825 mi

Los Angeles to Philadelphia: 2,743 mi

Houston to Detroit: 1,105 mi

NAME

Daily Review 73

Find each missing number. **Remember** to check your work. (Need an example? See pages 248–249.)

1. 5,250 mL = ________ L
2. 3.5 g = ________ mg
3. 8.2 L = ________ mL
4. 2.75 kg = ________ g
5. 650 g = ________ kg
6. 75 dL = ________ mL

Find each area. (Need an example? See pages 242–245.)

7. 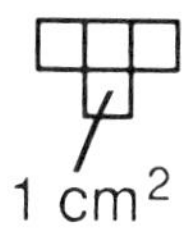

____

8. 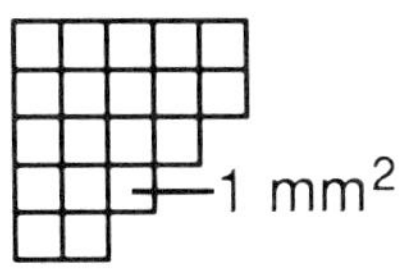

____

9. 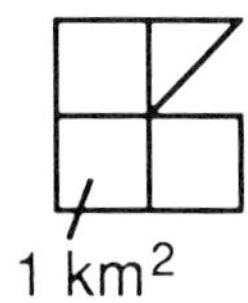

____

Under certain conditions, one milliliter of water weighs one gram and has a volume of one cubic centimeter. Use this information to find the following.
(Need an example? See pages 240–241.)

10. The volume of 250 mL of water ________
11. The weight of 250 mL of water ________
12. The weight of 2 L of water ________
13. The volume of 4 kg of water ________

NAME ____________________

Daily Review 74

Estimate each measure in the appropriate customary unit. (Need an example? See pages 250–251.)

1. 36 m ______ 2. 10 kg ______ 3. 4 L ______

4. 5 cm ______ 5. 5 km ______ 6. 20 cm ______

7. 500 g ______ 8. 7 L ______ 9. 200 g ______

Find each missing number. **Remember** to check your work. (Need an example? See pages 248–249.)

10. 42.3 L = ________ mL 11. 515 g = ________ kg

12. 6,125 mg = ________ g 13. 1,250 mL = ________ L

14. 125 dL = ________ L 15. 92 g = ________ mg

Find each volume. (Need an example? See pages 242–245.)

16.

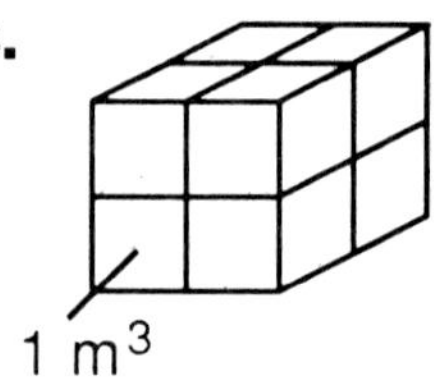

______

17.

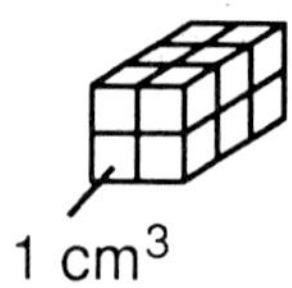

______

18.

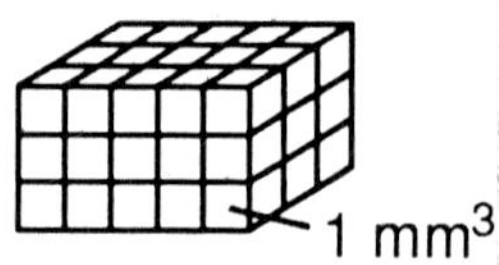

______

NAME

Daily Review 75

Give the unit of measure and the greatest possible error. (Need an example? See pages 252–253.)

1. 62 m ______________ **2.** $4\frac{1}{8}$ in. ______________

3. 93.7 km ______________ **4.** 2.75 cm ______________

5. $2\frac{1}{2}$ yd ______________ **6.** 35 mm ______________

Estimate each measure in the appropriate metric unit. (Need an example? See pages 250–251.)

7. 9 qt ________ **8.** 60 ft ________ **9.** 12 mi ________

10. 22 lb ________ **11.** 10 in. ________ **12.** 4 gal ________

13. 11 lb ________ **14.** 11 yd ________ **15.** 1.1 lb ________

Solve each problem. (Need an example? See pages 248–249.)

16. A 2-liter bottle of juice will fill how many 250-mL cups?

____

17. A 400-gram box of cereal contains 14 servings. About how many grams are in one serving?

____

18. A container holds 3,000 mL of water. What is its capacity in liters?

____

19. A cat weighs 4 kilograms. Find its weight in grams.

____

NAME

Daily Review 76

Add or subtract. **Remember** to rename minutes as hours in your answer when possible. (Need an example? See pages 254–255.)

1. 3 h 22 min + 2 h 55 min

2. 6 h 15 min + 4 h 20 min

3. 14 h 45 min + 7 h 25 min

4. 7 h 19 min − 3 h 4 min

5. 8 h 10 min − 2 h 26 min

6. 12 h − 5 h 11 min

Give the unit of measure and the greatest possible error.
(Need an example? See pages 252–253.)

7. 64 m ______________

8. $2\frac{1}{2}$ in. ______________

9. 12.3 mi ______________

10. 4.6 mm ______________

11. 35 cm ______________

12. 7.63 km ______________

Solve each problem. (Need an example? See pages 250–251.)

13. Joe has a 2-liter bottle of milk. About how many cups can he fill? ______________

14. Jill can walk 4 miles in one hour. About how many kilometers can she walk in one hour? ______________

NAME ____________

Daily Review 77

Write three equal ratios for each situation.
(Need an example? See pages 270–271.)

1. 6 pens to 7 pencils ____________

2. 16 test tubes to 4 Bunsen burners ____________

3. 12 jump ropes to 40 students ____________

4. 21 parents to 4 committees ____________

Find the elapsed time. (Need an example? See pages 254–255.)

5. From 6:25 A.M. to 8:10 A.M. ____________

6. From 8:20 A.M. to 1:05 P.M. ____________

7. From 7:45 A.M. to 6:20 P.M. ____________

8. From 10:50 A.M. to 4:35 P.M. ____________

Add or subtract. Round the answer to the less precise unit of measure. (Need an example? See pages 252–253.)

9. 5.02 m + 11.4 m ____________

10. $6\frac{3}{4}$ in. + 5 in. ____________

11. 23 cm + 31.2 cm ____________

12. 78 km − 16.1 km ____________

13. 11 yd − $2\frac{1}{3}$ yd ____________

14. 6.5 m − 2.05 m ____________

Use before pages 272–273.

NAME

Daily Review 78

Solve each proportion. (Need an example? See pages 272–273.)

1. $\frac{2}{3} = \frac{n}{15}$ ____________

2. $\frac{3}{5} = \frac{9}{x}$ ____________

3. $\frac{2}{7} = \frac{s}{21}$ ____________

4. $\frac{0.4}{3} = \frac{x}{24}$ ____________

5. $\frac{25}{30} = \frac{b}{12}$ ____________

6. $\frac{14}{35} = \frac{w}{30}$ ____________

Write four equal ratios. (Need an example? See pages 270–271.)

7. $\frac{1}{2}$ ____________________

8. $\frac{4}{7}$ ____________________

9. $\frac{20}{50}$ ____________________

10. $\frac{56}{7}$ ____________________

Solve each problem. Use the schedule at the right. (Need an example? See pages 254–255.)

11. How long will a trip take from Pittsburgh to Los Angeles on Flight 25?

____

12. How long will a trip take from Pittsburgh to Los Angeles on Flight 93?

____

From: Pittsburgh, PA
Eastern Standard Time
To: Los Angeles, CA
Pacific Standard Time

PITTS.	L. A.	Flight
7:45 A.M.	10:59 A.M.	93
9:55 A.M.	12:02 P.M.	25
2:30 P.M.	5:48 P.M.	11

NAME

Daily Review 79

Solve each proportion. (Need an example? See pages 272 and 274.)

1. $\frac{a}{4} = \frac{5}{2}$ ______

2. $\frac{0.4}{5} = \frac{x}{8}$ ______

3. $\frac{2.6}{8} = \frac{5.2}{y}$ ______

4. $\frac{c}{7} = \frac{2.4}{10}$ ______

5. $\frac{50}{6} = \frac{d}{9}$ ______

6. $\frac{25}{e} = \frac{10}{40}$ ______

7. $\frac{1}{4} = \frac{x}{28}$ ______

8. $\frac{4}{3} = \frac{y}{27}$ ______

9. $\frac{25}{10} = \frac{100}{z}$ ______

10. $\frac{15}{10} = \frac{x}{4}$ ______

11. $\frac{0.3}{5} = \frac{t}{15}$ ______

12. $\frac{12}{27} = \frac{20}{x}$ ______

Solve each problem. (Need an example? See pages 270–271.)

13. Velma has 25 classical music tapes and 13 rock music tapes. What is the ratio of classical music tapes to the total number of music tapes? ______

14. Elliot has a collection of 123 stamps. Of these, 20 stamps are from Europe. What is the ratio of European stamps to those that are not from Europe? ______

15. Martina has 150 baseball cards. Of these, 60 show American League players. What is the ratio of American League cards to those that are not American League? ______

NAME

Daily Review 80

Solve each problem by writing and solving a proportion. Round decimal answers to the nearest hundredth. (Need an example? See pages 276–277.)

1. Otis rode his bike 4.5 kilometers in 16 minutes. How far can he ride in 1 hour? ______

2. Oranges are selling at 3 for 89¢. How much will 24 oranges cost? ______

3. Jeanette drove 275 miles and her car used 14 gallons of gas. How far can she travel using 6 gallons of gas? ______

4. Workers in a factory produce 2,000 light bulbs in an 8-hour work day. How many light bulbs are produced in 2 hours? ______

Choose mental math, paper and pencil, or a calculator to solve each proportion. (Need an example? See pages 274–275.)

5. $\frac{24}{27}=\frac{48}{y}$ ______

6. $\frac{10}{12}=\frac{z}{30}$ ______

7. $\frac{0.2}{w}=\frac{1}{15}$ ______

Write a proportion, then solve.
(Need an example? See pages 272–273.)

8. Every 10 students in a pottery class need 3 pounds of clay. How many pounds of clay are needed for 25 students? ______

NAME

Daily Review 81

Is each pair of figures similar? Write *yes* or *no.* (Need an example? See pages 278–281.)

1.

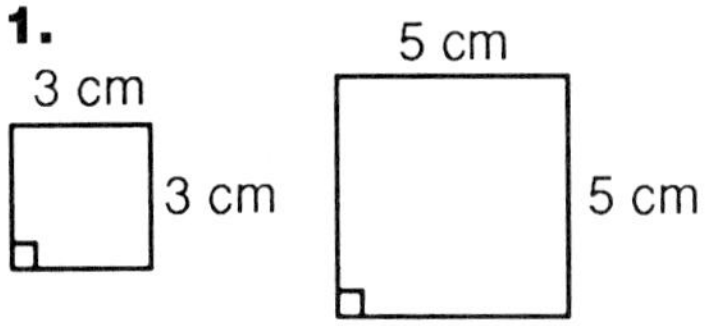

_____

2.

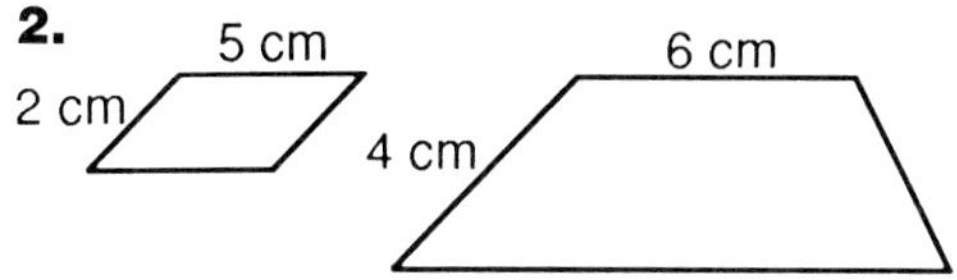

_____

3.

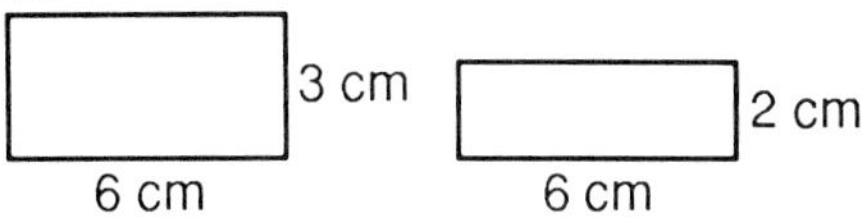

_____

4.

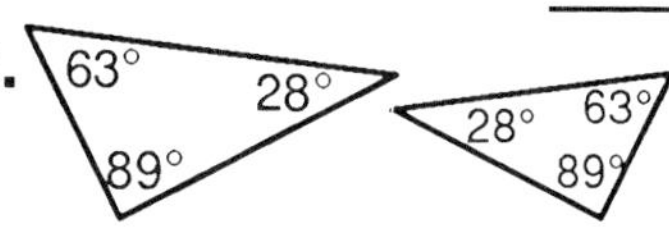

_____

Solve each problem by writing and solving a proportion.
(Need an example? See pages 276–277.)

5. Hank earns $4.75 per hour. How much will he earn if he works $6\frac{1}{2}$ hours? Round to the nearest cent.

_____

6. Ellen walked 5.2 km in 45 minutes. At that rate, about how long would it take her to walk 10 km?

_____

Write a proportion. Then solve.
(Need an example? See pages 274–275.)

7. A bookstore has fiction and nonfiction books in the ratio 4 to 5. If there are 400 nonfiction books, how many fiction books are there? _____

8. A recipe calls for 2 eggs to 3 cups of flour. How much flour is needed if 7 eggs are used? _____

NAME

Daily Review 82

Use the scale drawing at the right. Write a proportion to find the actual measures. Then solve. (Need an example? See pages 284–285.)

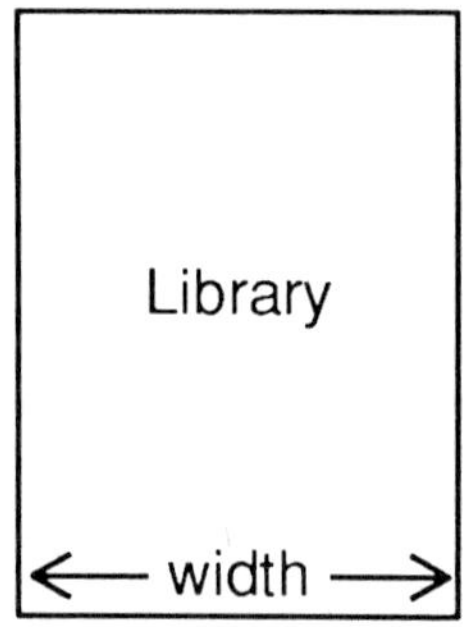

Scale: 1 cm = 3.5 m

1. What is the length of the library?

____

2. What is the width of the library?

____

Each pair of polygons below is similar. Find each missing length. (Need an example? See pages 278–281.)

3.

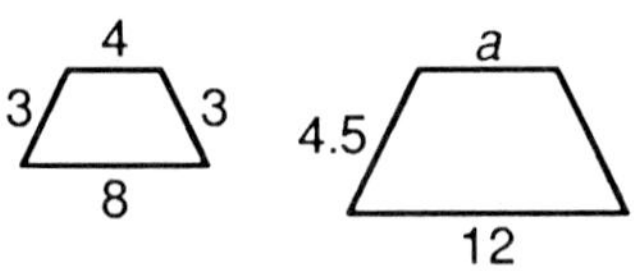

____

4.

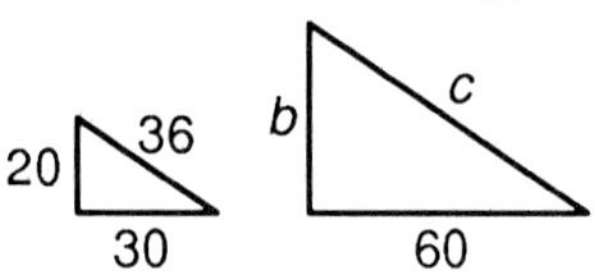

____

Solve each problem by writing and solving a proportion. (Need an example? See pages 276–277.)

5. If the exchange rate is 133 Japanese yen to 1 U.S. dollar, how many yen can be exchanged for $2,000?

____

6. If a car travels 250 miles on 12.5 gallons of gas, how many miles can the car travel on 10 gallons of gas?

____

Use before pages 286–287.

NAME ____________________

Daily Review 83

Solve each problem. Use the map at the right. (Need an example? See pages 286–287.)

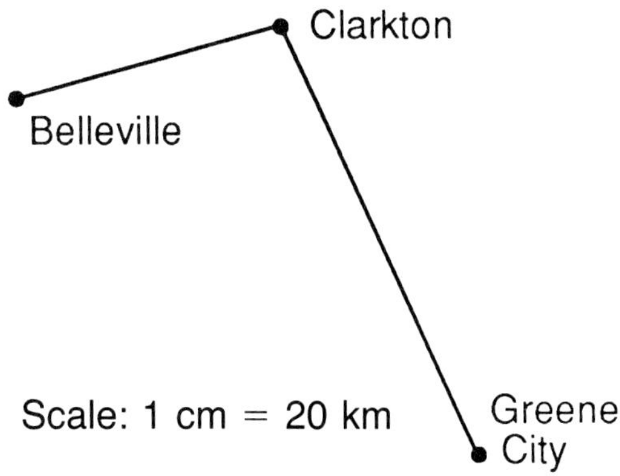

1. What is the distance between Belleville and Clarkton?

____

2. What is the distance between Clarkton and Greene City?

____

Use the scale drawing at the right. Write a proportion to find the actual measures. Then solve. (Need an example? See pages 284–285.)

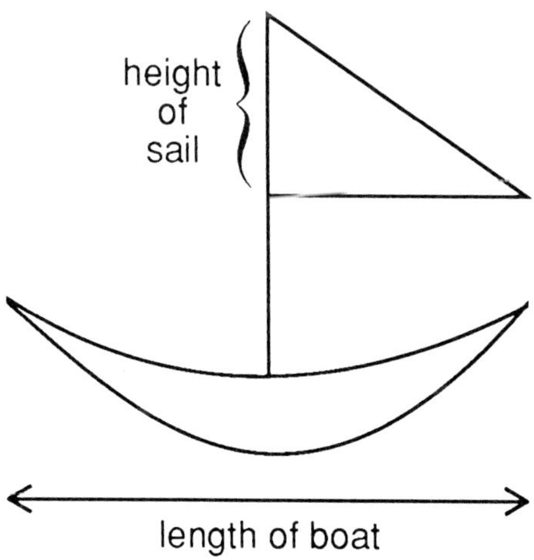

Scale: 1 cm = 2.5 m

3. What is the height of the sail? ____________

4. What is the length of the boat? ____________

NAME ______

Daily Review 84

Write each percent as a ratio. (Need an example? See pages 288–291.)

1. 47% ______ **2.** 29% ______ **3.** 2% ______

4. 25% ______ **5.** 6% ______ **6.** 11% ______

7. 98% ______ **8.** 4% ______ **9.** 34% ______

10. 1% ______ **11.** 10% ______ **12.** 100% ______

Solve the following problem. Use the map at the right. (Need an example? See pages 286–287.)

13. Sylvia rode her bike from Sleepy Hollow to Twin Oaks and from there to Breezy Point. How far did Sylvia travel?

______

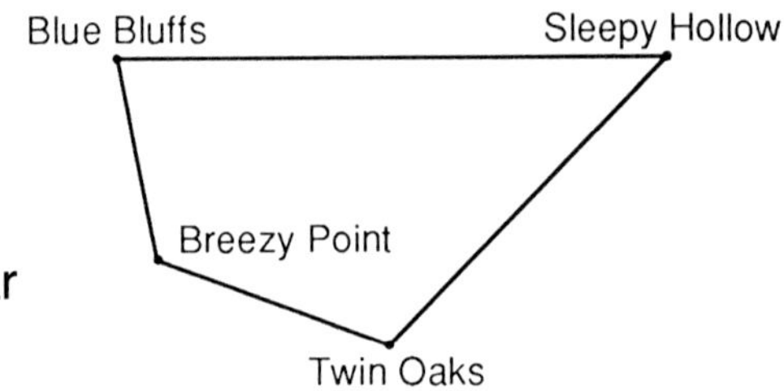

Scale: 1.5 cm = 6 km

Solve each problem. (Need an example? See pages 284–285.)

14. A model of a city has a scale of 1 yd to 2 mi. Two buildings are actually 6 mi apart. How far apart are they on the model?

______

15. On the model of the city, the harbor is $1\frac{1}{2}$ yd from City Hall. What is the actual distance between City Hall and the harbor?

______

NAME

Daily Review 85

Write each decimal as a percent. (Need an example? See pages 302–303.)

1. 0.27 ________ **2.** 0.48 ________ **3.** 0.07 ________

4. 0.2 ________ **5.** 0.8 ________ **6.** 2.11 ________

7. 0.125 ________ **8.** 0.009 ________ **9.** 0.024 ________

Write each ratio as a percent.
(Need an example? See pages 288–291.)

10. $\frac{23}{100}$ ________ **11.** $\frac{71}{100}$ ________ **12.** $\frac{3}{100}$ ________

13. $\frac{9}{10}$ ________ **14.** $\frac{3}{5}$ ________ **15.** $\frac{1}{2}$ ________

Solve each problem. Use the map at the right. (Need an example? See pages 286–287.)

16. How far is it from River Junction to Green Valley? Measure the map distance and then find the actual distance.

____

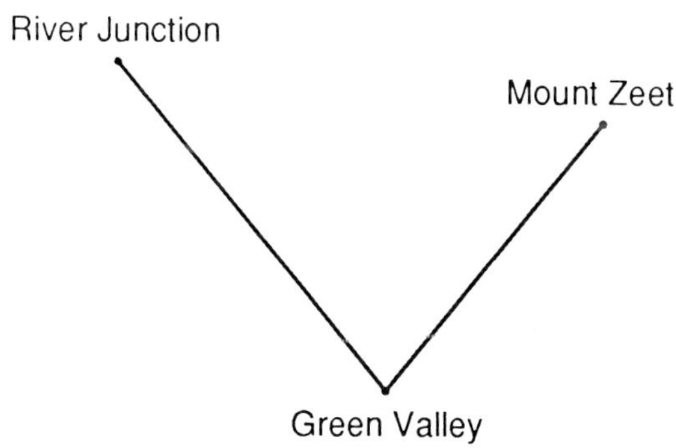

Scale: 2.2 cm = 10 km

17. How much farther is River Junction from Green Valley than is Mount Zeet from Green Valley? Round to the nearest tenth.

____

NAME

Daily Review 86

Write each fraction as a percent. (Need an example? See pages 304–307.)

1. $\frac{7}{100}$ ______ **2.** $\frac{17}{25}$ ______ **3.** $\frac{17}{20}$ ______

4. $\frac{1}{8}$ ______ **5.** $\frac{610}{100}$ ______ **6.** $\frac{0.3}{100}$ ______

7. $\frac{3}{4}$ ______ **8.** $\frac{20}{25}$ ______ **9.** $\frac{2}{3}$ ______

Write each percent as a decimal.
(Need an example? See pages 302–303.)

10. 21% ______ **11.** 73% ______ **12.** 2% ______

13. 11% ______ **14.** 60% ______ **15.** 50% ______

16. 5% ______ **17.** 99% ______ **18.** 80% ______

Write and solve a proportion for each. (Need an example? See pages 288–291.)

19. 12 is what percent of 50?

______

20. What is 20% of 70?

______

21. Find 75% of 60.

______

22. 25 is what percent of 200?

______

23. 10% of what number is 4?

______

24. 6 is 50% of what number?

______

NAME

Daily Review 87

Find each answer. Tell whether you wrote the percent as a fraction or a decimal.
(Need an example? See pages 308–311.)

1. 25% of 120 ________
2. 31% of 11 ________
3. 10% of 380 ________
4. 50% of 68 ________
5. $33\frac{1}{3}$% of 72 ________
6. 75% of 40 ________
7. 12% of 93 ________
8. 15% of 121 ________
9. 20% of 52 ________
10. 75% of 12 ________

Write each percent as a fraction or a mixed number. **Remember** to write the answer in lowest terms. (Need an example? See pages 304–307.)

11. 20% ______
12. 50% ______
13. 4% ______
14. 25% ______
15. 90% ______
16. 80% ______
17. 140% ______
18. 36% ______
19. 200% ______

Solve each problem. (Need an example? See pages 302–303.)

20. If 16 out of 100 people surveyed own a car, what percent of those surveyed own a car? ________
21. If 97 out of 100 people surveyed watch TV, what percent of the people surveyed watch TV? ________

NAME

Daily Review 88

Write an equation to find each answer.
(Need an example? See pages 314–315.)

1. What percent of 5 is 4? ____________

2. What percent of 8 is 4? ____________

3. 6 is what percent of 20? ____________

4. 10 is what percent of 5? ____________

5. What percent of 60 is 15? ____________

6. What percent of 250 is 10? ____________

For Exercises 7–10, tell whether you would use mental math, paper and pencil, or a calculator. Then find each answer. (Need an example? See pages 308–311.)

7. 50% of 520 ____________

8. 25% of 36 ____________

9. 10.3% of 231 ____________

10. 200% of 156 ____________

Complete the table. (Need an example? See pages 304–307.)

	Fraction	Decimal	Percent
11.	$\frac{3}{4}$		
12.		0.15	
13.			60%

NAME

Daily Review 89

Use an equation to find each number.
(Need an example? See pages 316–317.)

1. 75% of what number is 150?

2. 16 is 50% of what number?

3. 42 is 10% of what number?

4. 20% of what number is 11?

5. 150% of what number is 6?

6. $\frac{1}{2}$% of what number is 2.4?

Find each answer. (Need an example? See pages 314–315.)

7. What percent of 10 is 7?

8. What percent of 9 is 81?

9. 6 is what percent of 50?

10. 120 is what percent of 90?

Solve each problem. (Need an example? See pages 308–311.)

11. A factory produced 70,000 machine parts. The quality control inspector reported that 4% were defective. How many parts were defective?

12. Joe's hourly rate is 150% of what it was 5 years ago. He made $18 per hour five years ago. What is his hourly wage now?

NAME

Daily Review 90

Tell whether you would use paper and pencil or a calculator. Then solve each problem. (Need an example? See pages 318–319.)

1. Winston invested $600 at 5.8% simple interest for 3 years. How much interest did he earn?

Method: ______________

Answer: ______________

2. Julia borrowed $1,000 at 12% simple interest for 1 year. How much interest did she pay?

Method: ______________

Answer: ______________

Use an equation to find each number. (Need an example? See pages 316–317.)

3. 18 is 25% of what number?

____

4. 300% of what number is 27?

____

5. $\frac{1}{2}$% of what number is 10?

____

6. 60% of what number is 45?

____

Solve each problem. (Need an example? See pages 314–315.)

7. There are 250 travel books in the library. Of these, 80 are about tropical islands. What percent of the travel books are about tropical islands? ______________

8. Of the 675 junior high school students, 450 take a school bus. What percent of the students take a school bus? ______________

NAME

Daily Review 91

Estimate.
(Need an example? See pages 320–321.)

1. 75% of 15,749

2. 23% of 412

3. 11% of what number is 18?

4. What percent of 61 is 29?

5. What percent of 515 is 98?

6. 47% of what number is 82?

Tell whether you would use paper and pencil or a calculator. Then solve each problem. (Need an example? See pages 318–319.)

7. Maria invested $450 at $7\frac{1}{2}$% simple interest 2 years ago. How much interest has she earned?

8. Ted borrowed $275 at 14% simple interest for 6 months. How much interest did he pay?

Solve each problem. (Need an example? See pages 316–317.)

9. Mrs. Russo earned $1,500 in commission selling computer chips. This was 3% of her total sales. Find her total sales.

10. Geraldine needs a 20% deposit on the used car she wants to buy. The deposit is $640. What is the price of the car?

NAME

Daily Review 92

Write and solve an equation to answer Exercises 1–6.
(Need an example? See pages 322–323.)

1. 15% of what number is 45? __________

2. 56 is 80% of what number? __________

3. What percent of 96 is 12? __________

4. What percent of 120 is 4? __________

5. What percent of 50 is 125? __________

6. 250% of what number is 20? __________

Estimate. (Need an example? See pages 320–321.)

7. 63% of 732 __________

8. 24% of 15,782 __________

9. 9% of what number is 29? __________

10. What percent of 42 is 9? __________

Solve each problem. Tell whether you used paper and pencil or a calculator. (Need an example? See pages 318–319.)

11. Louis invested \$10,000 at $8\frac{1}{2}$% simple interest for 4 years. How much interest did he earn? __________

12. Althea borrowed \$900 for 2 years at 15% simple interest. How much interest did she pay? __________

NAME

Daily Review 93

Tell whether you would use paper and pencil or a calculator. Then solve each problem. (Need an example? See pages 324–325.)

1. A computer is selling for 20% off the regular price. How much does the computer cost if its regular price is $2,500?

____

2. Jill bought a software package that cost $349. What was her total cost if the sales tax was 3%?

____

Write and solve an equation to answer Exercises 3–8. (Need an example? See pages 322–323.)

3. 25% of what number is 11?

____

4. What percent of 90 is 18?

____

5. What percent of 200 is 25?

____

6. 60% of what number is 180?

____

Estimate. (Need an example? See pages 320–321.)

7. 51% of 1,625

____

8. What percent of 310 is 92?

____

9. 67% of what number is 48?

____

10. 23% of 782

____

NAME

Daily Review 94

Tell whether each figure is a prism. Explain your answer. (Need an example? See pages 336–339.)

1.

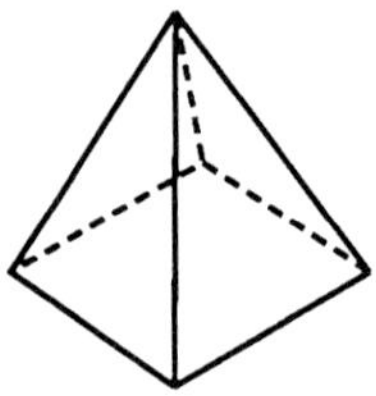

2.

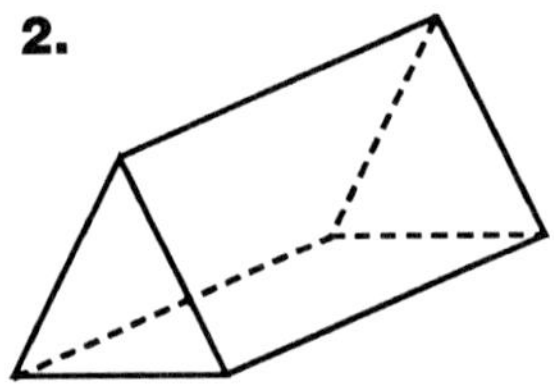

3.

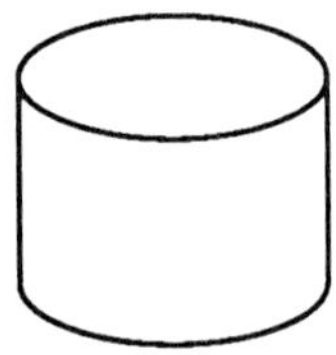

Tell whether you would use pencil and paper or a calculator. Then solve each problem. (Need an example? See pages 324–325.)

4. A book is on sale for \$15. The original price was \$24. What is the percent of the discount?

5. A coat is on sale for 30% of the regular price. The original price was \$80. What is the sale price?

Write and solve an equation to answer Problems 6 and 7. Round to the nearest whole number. (Need an example? See pages 322–323.)

6. A baseball team won 54% out of 80 games. How many games did they win?

7. Fred made 85% of the field goals he attempted. He made 34 goals. How many did he attempt?

NAME

Daily Review 95

Use the diagrams below to help you find a pattern and then use it to complete the table. (Need an example? See pages 340–341.)

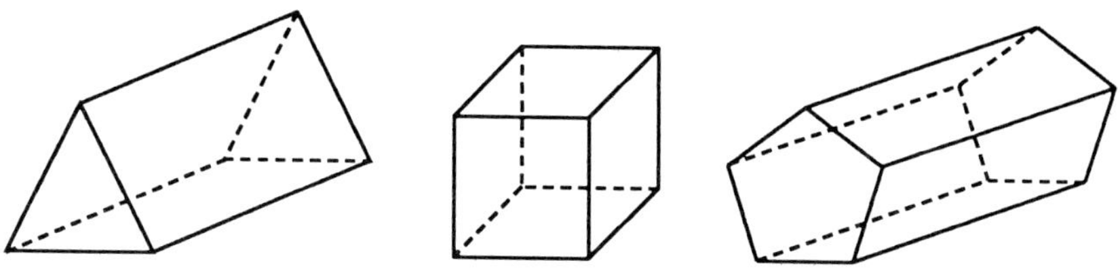

Number of sides in the base of the prism	3	4	5	10	20
Number of vertices in the prism	**1.**	**2.**	**3.**	**4.**	**5.**

Name the shaded part of each polyhedron. (Need an example? See pages 336–339.)

6.

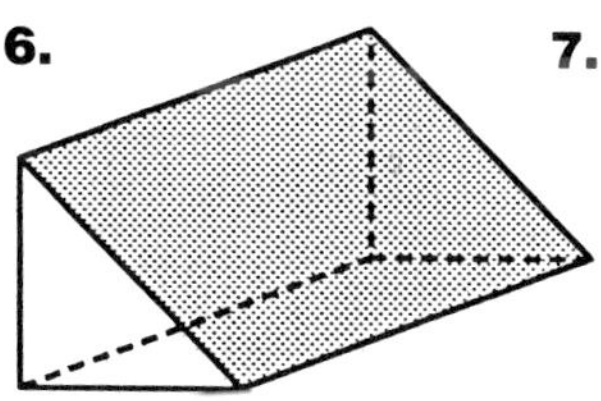

7.

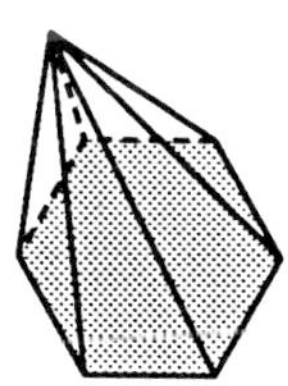

8.

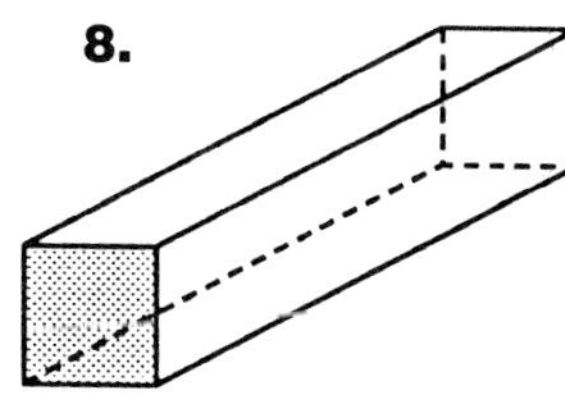

Solve the problem. (Need an example? See pages 324–325.)

9. Sam bought an album for $8.99. He paid 4% sales tax. What was the total cost of the album? Round to the nearest cent.

NAME ______

Daily Review 96

Find each perimeter and area. (Need an example? See pages 342–345.)

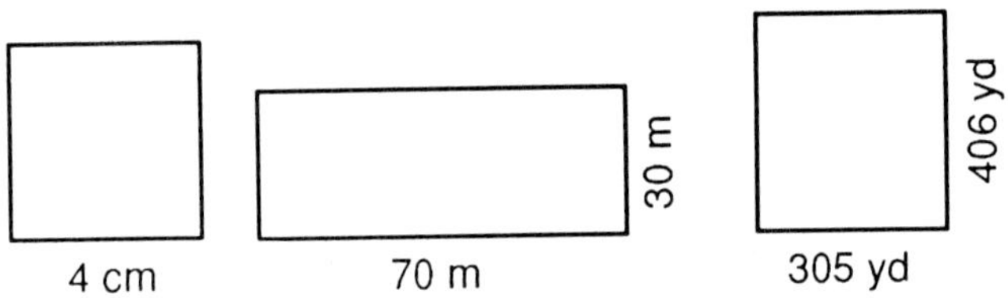

Perimeter: **1.** ______ **2.** ______ **3.** ______

Area: ______ ______ ______

Use the diagram at the right to find a pattern. Use it to complete Exercises 4 and 5. (Need an example? See pages 340–341.)

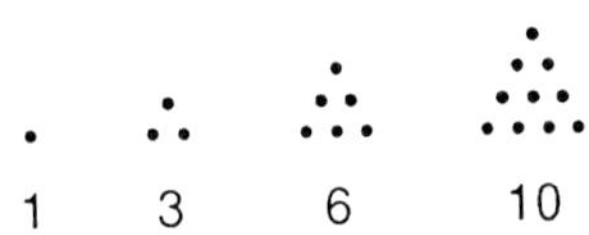

4. The diagram shows triangular numbers. What is the tenth triangular number?

______

5. What is the 20th triangular number?

______

Complete the table. (Need an example? See pages 336–339.)

	Number of	
6.	Faces	
7.	Vertices	
8.	Edges	

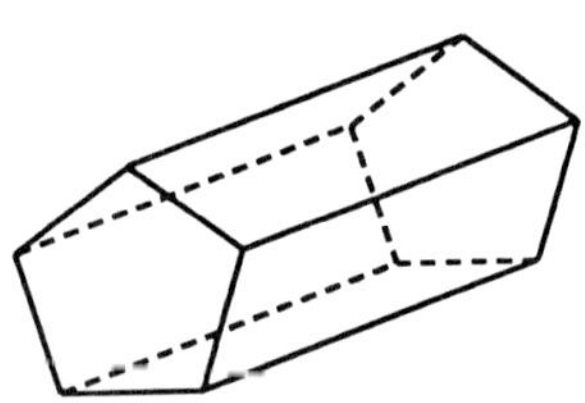

NAME

Daily Review 97

Find the perimeter and the area of each figure. **Remember** that the height is perpendicular to the base. (Need an example? See pages 346–347.)

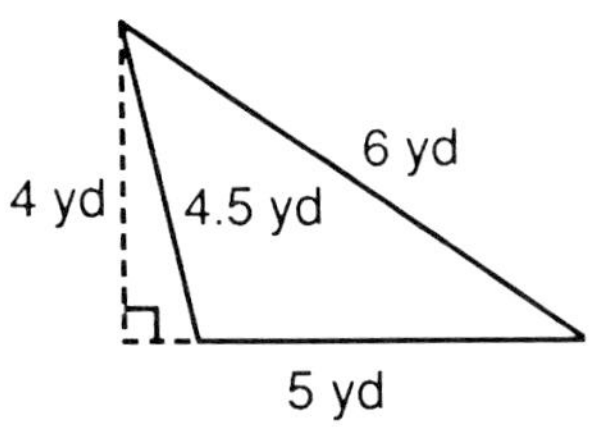

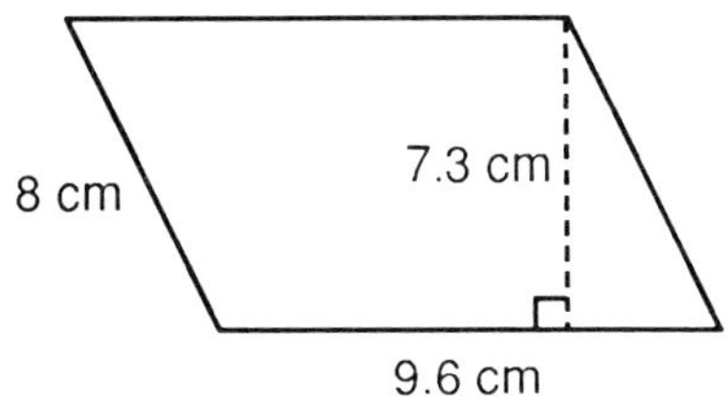

Perimeter: **1.** ________ **2.** ________

Area: ________ ________

Find the area and the perimeter. (Need an example? See pages 342–345.)

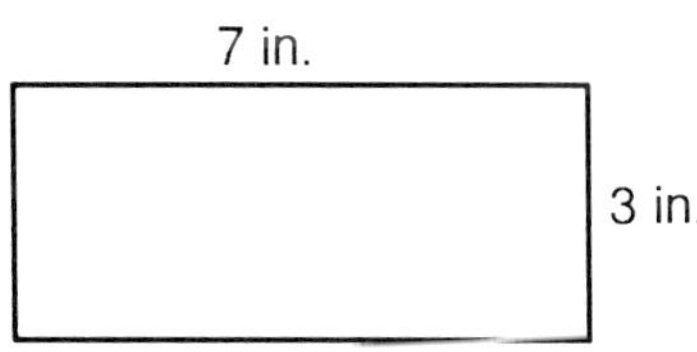

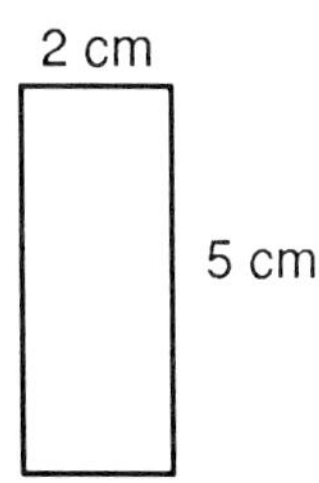

Perimeter: **3.** ________ **4.** ________

Area: ________ ________

 Use before pages 348–349.

NAME

Daily Review 98

Find the perimeter and the area of each figure. (Need an example? See pages 348–349.)

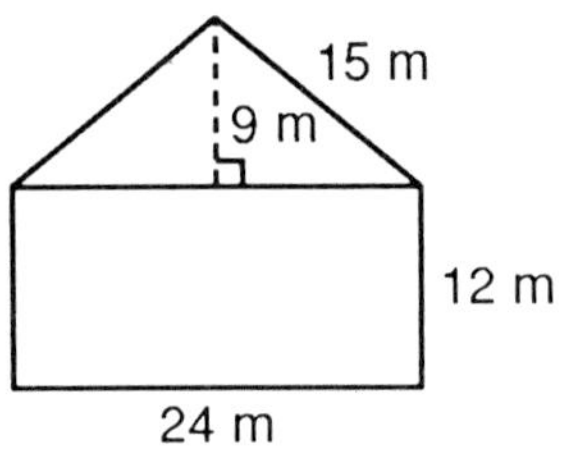

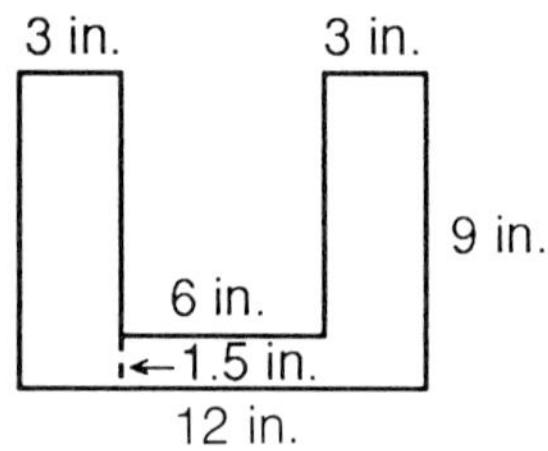

Perimeter: **1.** ____________ **2.** ____________

Area: ____________ ____________

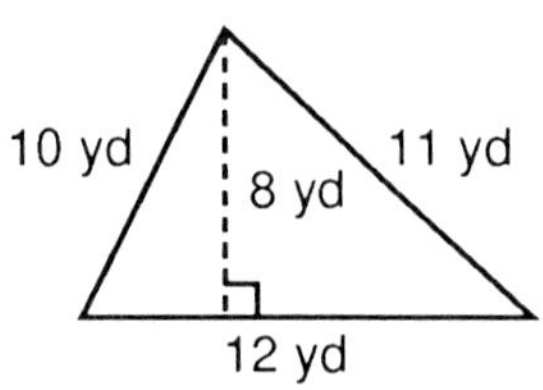

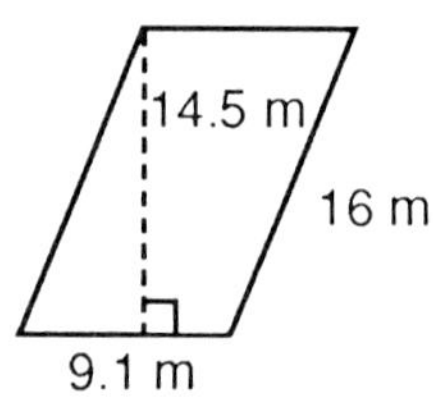

Perimeter: **3.** ____________ **4.** ____________

Area: ____________ ____________

Find the area and the perimeter of each rectangle. Tell whether you use mental math or paper and pencil. (Need an example? See pages 342–345.)

	$l = 16$ in.; $w = 14$ in.	$s = 70$ cm	$l = 5\frac{1}{2}$ ft; $w = 2$ ft
Perimeter:	**5.** ____________	**6.** ____________	**7.** ____________
Area:	____________	____________	____________

NAME

Daily Review 99

Find the circumference. Use 3.14 for π.
(Need an example? See pages 350–351.)

1. $r = 11$ mm ________ **2.** $d = 21$ cm ________ **3.** $d = 100$ ft ________

Find the perimeter and the area of each polygon. (Need an example? See pages 346 and 348.)

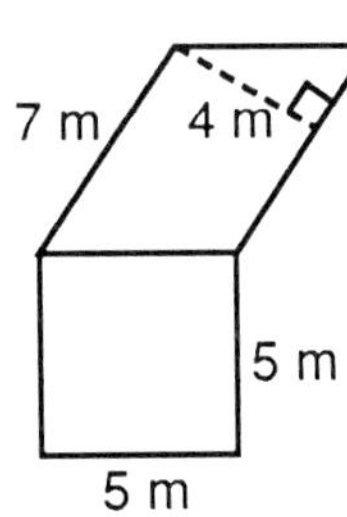

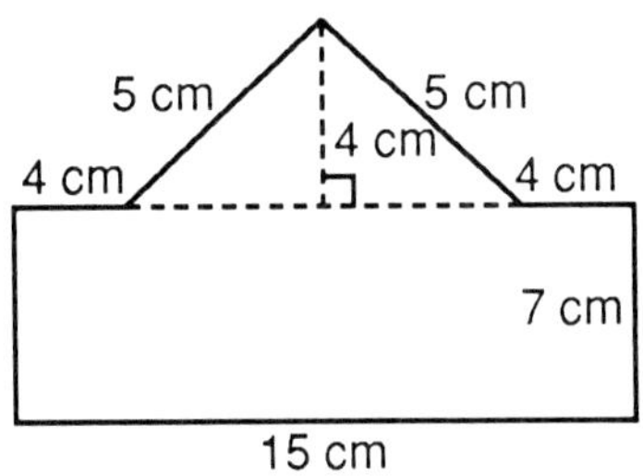

Perimeter: **4.** ________ **5.** ________

Area: ________ ________

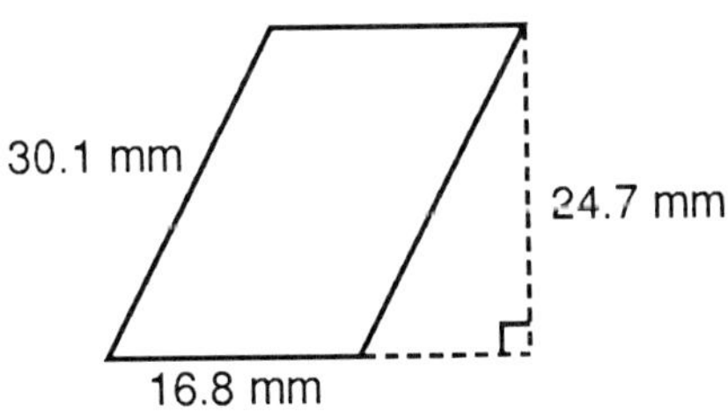

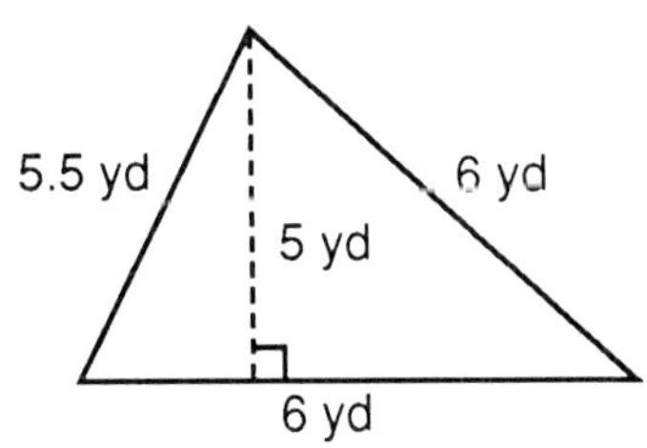

Perimeter: **6.** ________ **7.** ________

Area: ________ ________

 Use before pages 352–354.

NAME

Daily Review 100

Find the area of each circle. Use 3.14 for π. **Remember** to label units in your answer. (Need an example? See pages 352–354.)

1. Diameter: 20 mm
2. Radius: 22.5 cm
3. Diameter: 100 m
4. Radius: 78 cm
5. Diameter: 200 yd
6. Radius: 0.1 m

Find the circumference. Use 3.14 for π. (Need an example? See pages 350–351.)

7. Radius: 4 ft
8. Diameter: 36.2 cm
9. Radius: 10.5 in.
10. Diameter: 1.9 m

Solve each problem. (Need an example? See pages 342–345.)

11. A rectangular courtyard has a length of 100 m and a width of 50 m. Find the perimeter and the area of the courtyard.
12. A square window has a perimeter of 120 cm. Find the length of each side and its area.

NAME

Daily Review 101

Find the surface area of each polyhedron.
(Need an example? See pages 358–359.)

1.

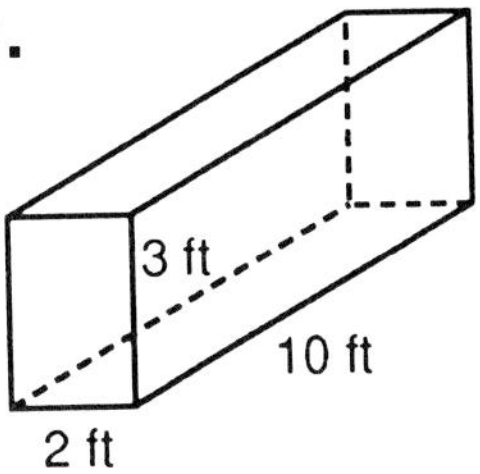

2.

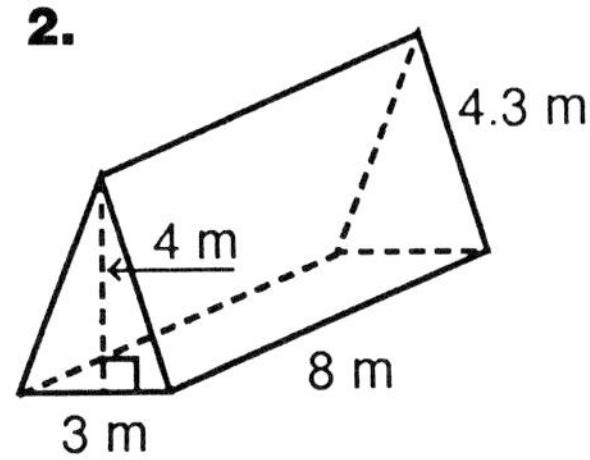

3.

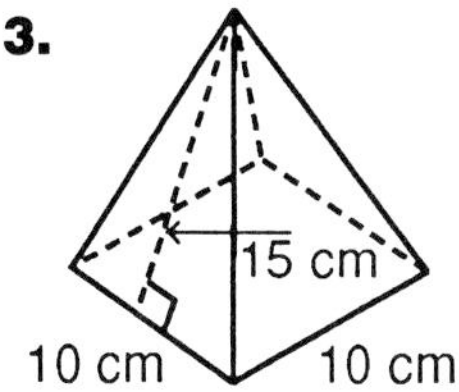

__________ __________ __________

Find the area of each circle. Use $\frac{22}{7}$ for π. (Need an example? See pages 352–354.)

4. Radius: $24\frac{1}{2}$ ft **5.** Diameter: $2\frac{1}{3}$ yd **6.** Radius: $10\frac{1}{2}$ yd

__________ __________ __________

7. Diameter: $1\frac{1}{6}$ ft **8.** Radius: 42 in. **9.** Diameter: 14 in.

__________ __________ __________

Solve each problem. Use 3.14 for π. (Need an example? See pages 350–351.)

10. A circular garden has a radius of 60 m. How many meters of fencing are needed to go around it?

____

11. Jane wants to wrap ribbon around a circular frame. The frame has a diameter of 15 cm. How much ribbon does she need?

____

NAME

Daily Review 102

Find the surface area of each cylinder. Use 3.14 for π. **Remember** to include both bases in your calculations. (Need an example? See pages 360–361.)

1. A cylinder with a radius of 10 mm and a height of 20 mm

____

2. A cylinder with a diameter of 6 cm and a height of 12 cm

____

Find the surface area of each polyhedron. (Need an example? See pages 358–359.)

3.

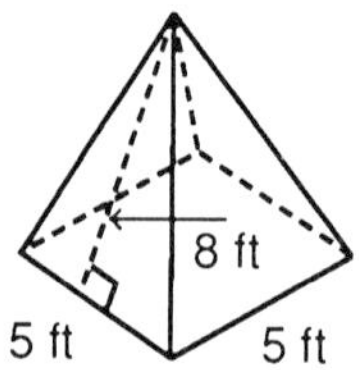

____

4.

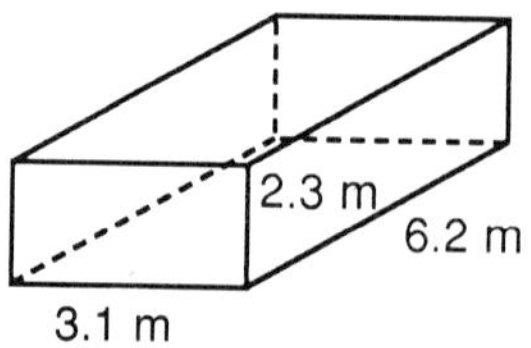

____

5.

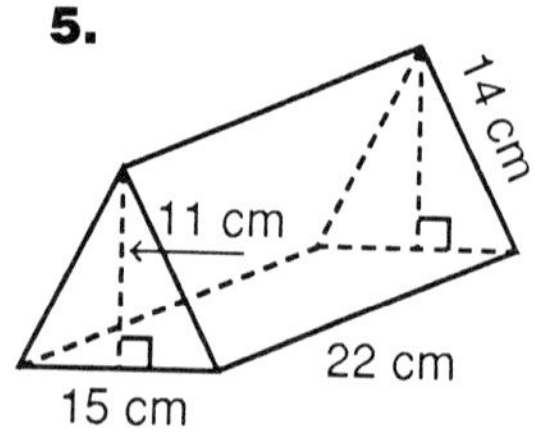

____

Find the area of each shaded region. Use 3.14 for π.
(Need an example? See pages 352–354.)

6.

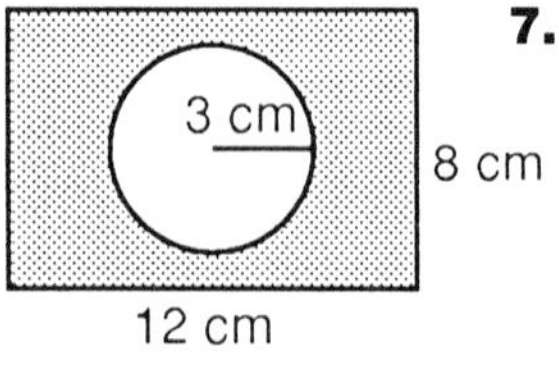

____

7.

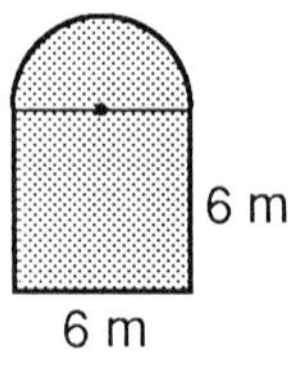

____

8.

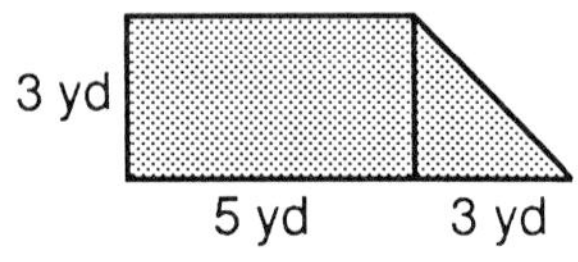

____

NAME ____________________

Daily Review 103

Find each volume. Use 3.14 for π. (Need an example? See pages 362–363.)

1.

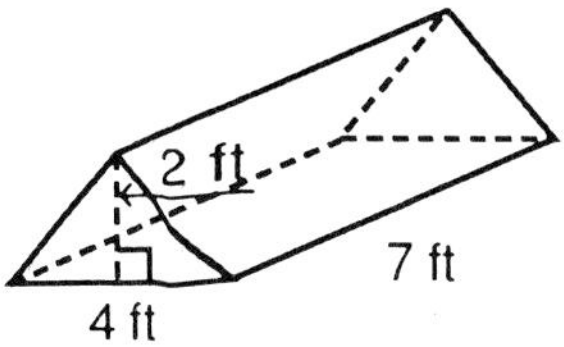

2.

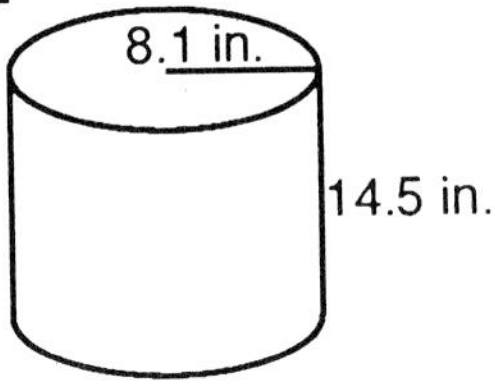

3.

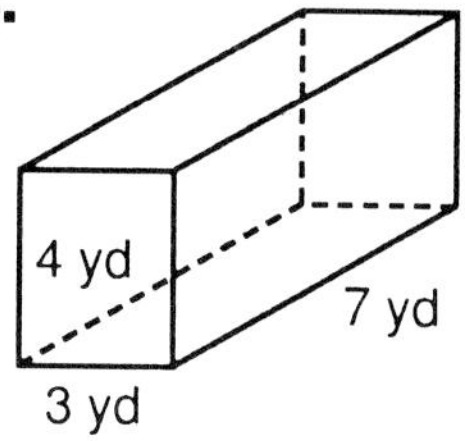

Find the surface area of each cylinder. Use 3.14 for π. (Need an example? See pages 360–361.)

4.

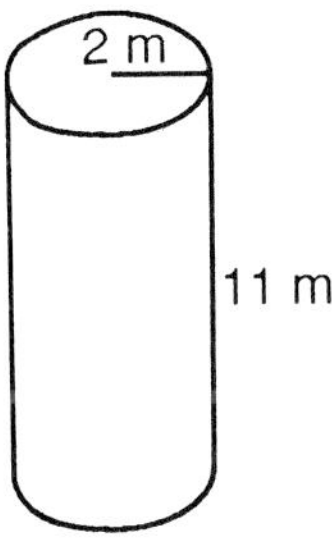

5.

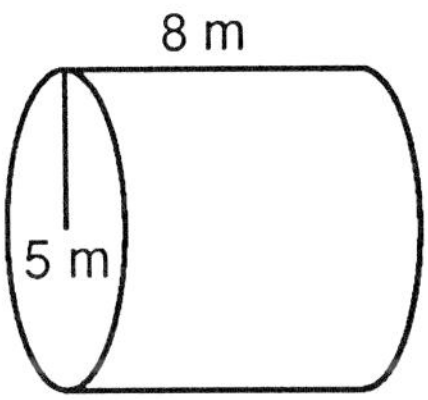

6.

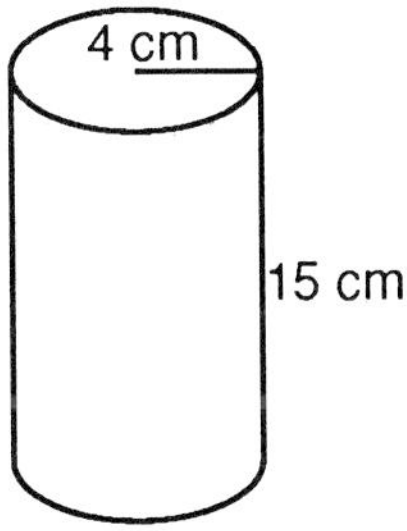

Solve each problem. (Need an example? See pages 358–359.)

7. A room is 4 m long, 3 m wide and 2.5 m high. What is the surface area of the four walls and the ceiling?

8. Carpeting costs $12 per square meter. What will it cost to carpet the floor of this room?

NAME ____________________

Daily Review 104

Find the volume of each figure to the nearest whole number. Use 3.14 for π.
(Need an example? See pages 364–365.)

1.

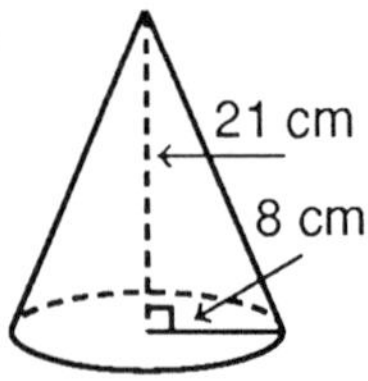

2.

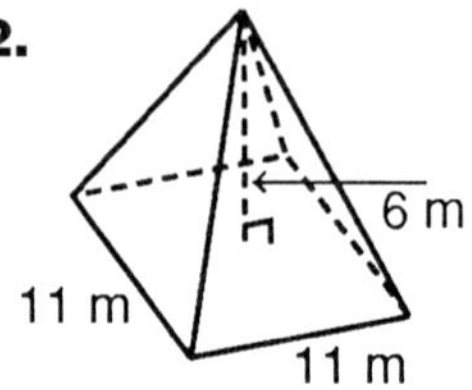

3.

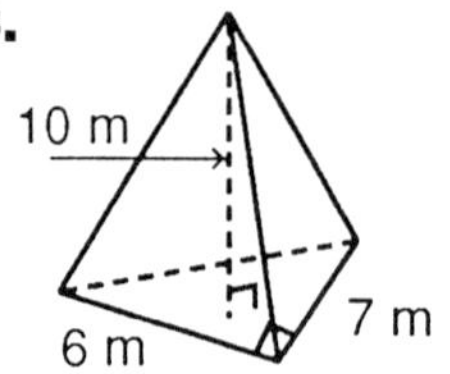

______ ______ ______

Find each volume. Use 3.14 for π. (Need an example? See pages 362–363.)

4.

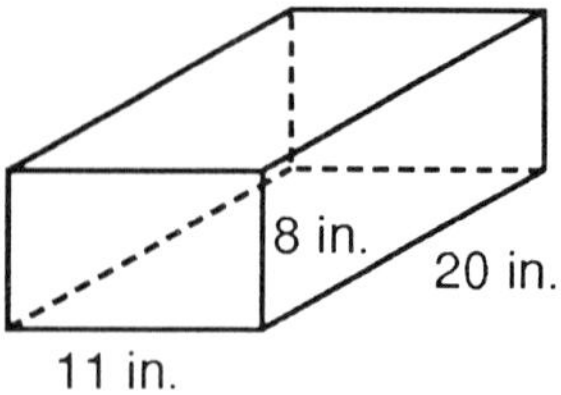

5.

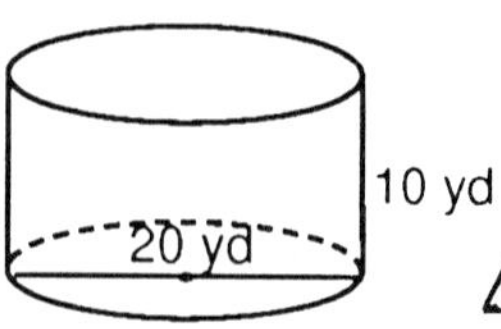

6.

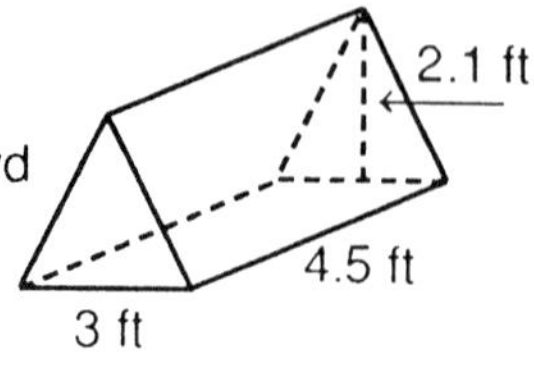

______ ______ ______

Solve each problem. Use 3.14 for π. (Need an example? See pages 360–361.)

7. A can of soup has a height of 12 cm and a base radius of 5 cm. What is the surface area of its label (not including its two bases)?

______

8. A mailing tube has a height of 50 cm and a radius of 10 cm. What is its surface area?

______

NAME

Daily Review 105

Solve each problem. (Need an example? See pages 366–367.)

1. Find the circumference and the area of the two circles shown below.

____

____

2. How many times greater is the area of circle B? How many times greater is the circumference of circle B?

____

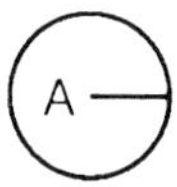

r = 3 cm

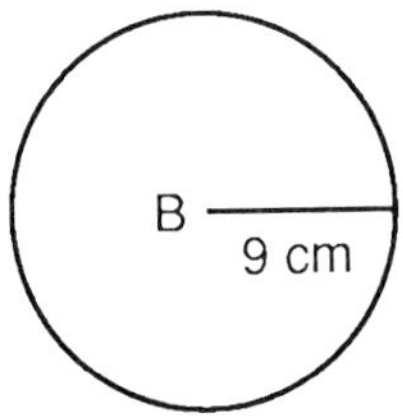

Find the volume of each figure to the nearest whole number. Use 3.14 for π. (Need an example? See pages 364–365.)

3.

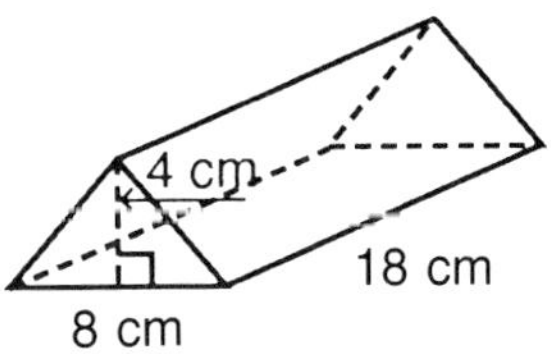

4.

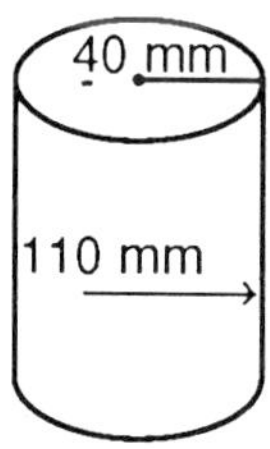

5.

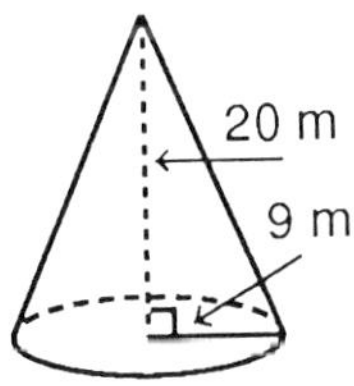

_______________ _______________ _______________

Solve the problem. (Need an example? See pages 362–363.)

6. A cylindrical trash basket has a base radius of 5 in. and a height of 18 in. A rectangular trash basket measures 12 in. by 6 in. by 18 in. Which holds more?

____

NAME ______________________

Daily Review 106

Solve each problem. Use 3.14 for π and round answers to the nearest tenth. (Need an example? See pages 368–369.)

1. How much popcorn can the cone hold?

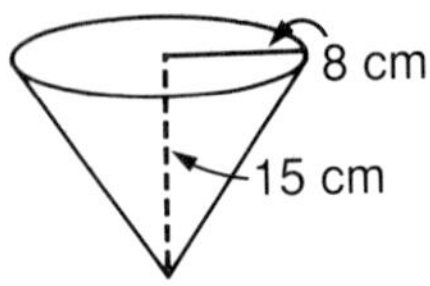

____

2. What is the volume of the trunk?

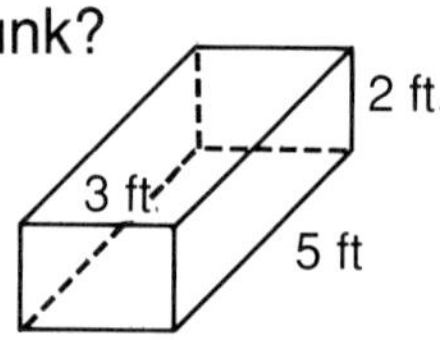

____

Answer these questions about the cube shown. (Need an example? See pages 366–367.)

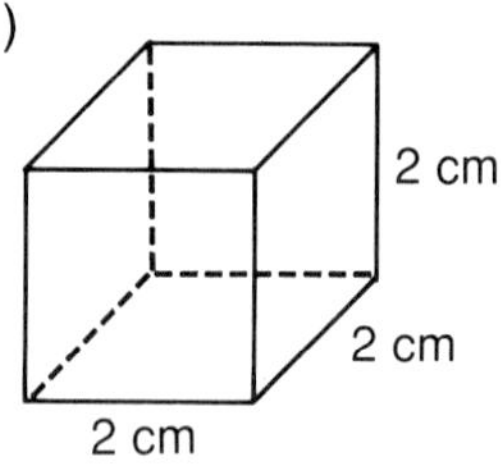

3. What is the volume of a new solid formed when one side is tripled?

____

4. What is the volume when the length and width are tripled?

____

5. What is the volume when all three dimensions are tripled?

____

Solve the problem. Round to the nearest tenth. (Need an example? See pages 364–365.)

6. A cone has a height of 10 cm, and its radius is 5 cm. How much sand does it hold?

____

NAME

Daily Review 107

Tell whether you would use a calculator or paper and pencil. Then find the mean of each set of data. Round to the nearest tenth. (Need an example? See pages 380–383.)

1. 6 8 7 9 ____________

2. 20 31 19 28 ____________

3. 624 432 659 501
785 641 590 578 ____________

4. 42 51 68 78 92
63 69 72 58 49
84 65 73 52 81 ____________

Solve each problem. Use 3.14 for π and round answers to the nearest tenth. (Need an example? See pages 368–369.)

5. Find the area of the elliptical table top. ____________

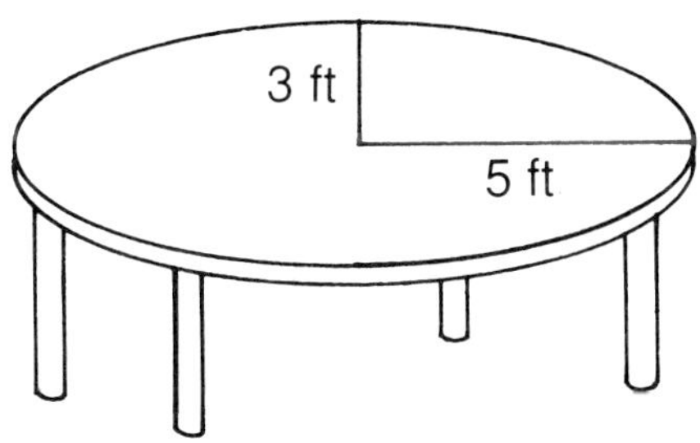

6. The circumference of a round table is 5.6 m. What is its radius? ____________

Solve the problem. (Need an example? See pages 366–367.)

7. If each dimension of this rectangle is doubled, what will be its area? ____________

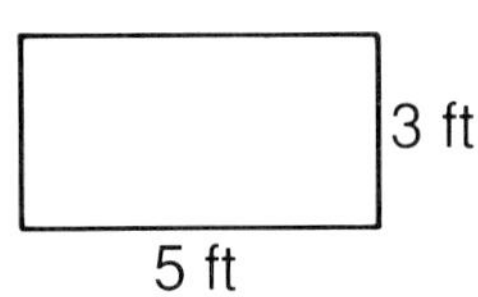

NAME

Daily Review 108

Nick asked 15 friends how many television programs they watched in a day. Use the results of his survey to answer Exercises 1–2. (Need an example? See pages 384–387.)

3 1 0 4 0 5 1 6 2 4 1 1 4 3 6

1. Complete the frequency table.

Number of Shows	Tally	Frequency

2. Make a line plot of the data.

Find the mode for each set of data. (Need an example? See pages 380–383.)

3. 245 316 289 245
267 289 302 289

____

4. 56 75 39 42 40
61 55 40 72 36
64 73 82 51 45

____

Solve the problem. Use 3.14 for π. (Need an example? See pages 368–369.)

5. Find the volume of a cylindrical tank that has a base radius of 15 m and a height of 50 m. ______________

NAME

Daily Review 109

Answer each question. (Need an example? See pages 388–389.)

1. A total of 500 raccoons were captured, tagged, and released back into a forest preserve. Then a sample of 100 were captured, 40 of which had been tagged. How many raccoons are in the forest?

   ____

2. Four thousand residents from Bedford commute to a nearby city. When 100 Bedford residents were surveyed, 25 said that they were commuters. What is the population of Bedford?

   ____

The following data show the number of after-school activities that 19 seventh-grade students participated in during one week. Use the data to answer Exercises 3–4. (Need an example? See pages 384–387.)

1 0 2 3 2 2 3 0 0 1 4 2 2 3 0 2 1 1 2

3. Complete the frequency table.

Number of Activities	Tally	Frequency

4. Make a line plot of the data.

NAME

Daily Review 110

Make a stem-and-leaf plot for each set of data. (Need an example? See pages 390–391.)

1. 16 14 25 18 32 15
36 24 19 21 28 34

2. 89 115 92 86 110
105 95 88 96 101

3. 86 70 63 77 61 72
61 72 89 85 64 75

4. 43 59 38 50 36 45
36 48 36 41 59 54

Answer each question. (Need an example? See pages 388–389.)

5. The health club in Belville has 300 members. In a survey of 100 Belville residents, 20 said that they belonged to the health club. What is the population of Belville?

____

6. A total of 600 squirrels were captured, tagged, and released back into a woodland. A sample of 200 captured squirrels included 50 that were tagged. How many squirrels are in the woodland?

____

NAME

Daily Review 111

Use the data in the table for Exercise 1. Draw the graph on a separate sheet of paper. (Need an example? See pages 394–395.)

1. Make a bar graph showing the number of books each student read this week.

Student	Number of Books
Michelle Z.	3
Chris J.	4
Anthony B.	3
Elena R.	3
Fred A.	2
Abby K.	4

Make a stem-and-leaf plot for each set of data. (Need an example? See pages 390–391.)

2. 56 82 71 63 65
78 61 75 59 62

3. 25 28 32 41 38
29 35 46 37 42

Answer the question. (Need an example? See pages 388–389.)

4. A total of 150 starlings were captured, tagged, and released back into a state park. A sample of 50 captured starlings included 10 that were tagged. How many starlings are in the state park? ____________

  Use before pages 396–397.

NAME

Daily Review 112

Use the circle graph showing the Lewis' spending last month. (Need an example? See pages 396–397.)

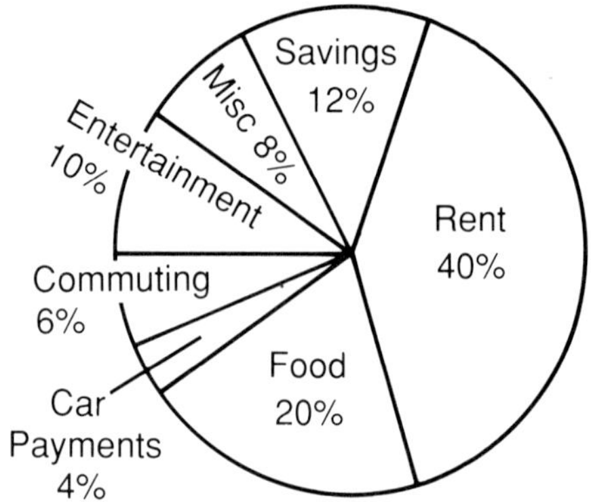

1. What percent of their spending last month was for commuting and car payments? _______

2. Which two categories accounted for exactly 60% of the total?

   ____

3. For which category was the amount 12% of the total?

   ____

Use the data in the table for Exercises 4–5. Draw graphs on a separate sheet of paper.
(Need an example? See pages 394–395.)

Books Read This Month

Number of books	0	1	2	3	4	5	6	7
Frequency (7th graders)	2	4	5	3	8	9	7	5
Frequency (8th graders)	3	3	4	5	6	10	8	6

4. Make a double bar graph comparing the number of books read by 7th and 8th graders this month.

5. Make a histogram of the data showing the number of books read by 7th graders. Use intervals of two books.

NAME

Daily Review 113

Use the data in the chart for Exercises 1–2. Draw graphs on a separate sheet of paper. (Need an example? See pages 398–399.)

Population (in thousands)

Year	Hilldale	Newton
1900	13	20
1920	20	40
1940	22	36
1960	28	32
1980	32	26

1. Draw a broken-line graph for the population of Hilldale.
2. Graph the data for Newton on the same grid to make a double broken-line graph.

Use the circle graph showing the Ace Sporting Goods sales breakdown for last year. If the total sales were $400,000, how much did they sell in each category? (Need an example? See pages 396–397.)

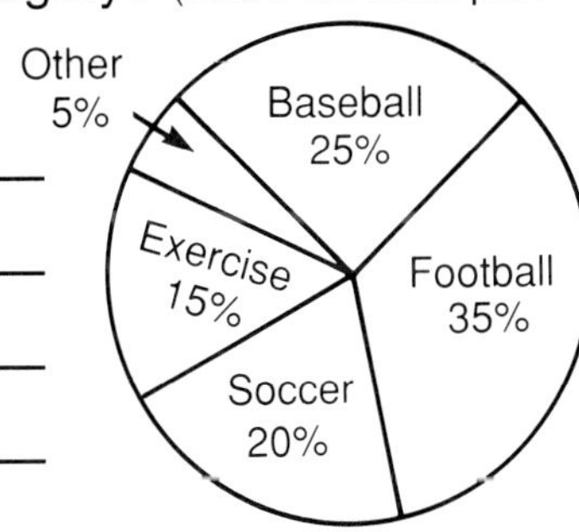

3. Baseball equipment ____________
4. Exercise equipment ____________
5. Football equipment ____________
6. Soccer equipment ____________

Use the data in this table for Exercise 7. (Need an example? See pages 394–395.)

Student	Jill	Otis	Nick	Elyse	Vera	Peter
Min. exercise	15	30	15	45	30	15

7. On a separate paper, make a bar graph showing the number of minutes per day each student exercises.

NAME ______________________

Daily Review 114

The table shows the number of points scored in each quarter by two basketball teams. Tell which kind of graph you would make to help you answer each question. Then answer the questions. Assume you have collected all the necessary data. (Need an example? See pages 400–401.)

Points Scored

Quarter	Reds	Blues
1st	28	20
2nd	15	22
3rd	19	28
4th	10	6

1. Which team scored more points in the third quarter?

____

2. For which team did the number of points decrease from the first to the second quarter?

____

Use the data in the chart for Exercises 3–4. Draw graphs on a separate sheet of paper. (Need an example? See pages 398–399.)

3. Draw a broken-line graph for the number of T-shirts sold each week by the seventh graders.

4. Graph the data for the eighth graders on the same grid to make a double broken-line graph.

Number of T-shirts Sold

Week	7th Graders	8th Graders
1	10	15
2	9	12
3	11	6
4	8	4

NAME

Use these two bar graphs for Exercises 1–3. (Need an example? See pages 402–403.)

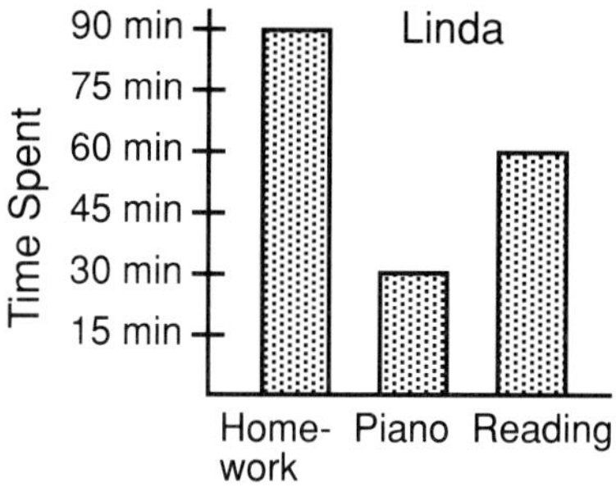

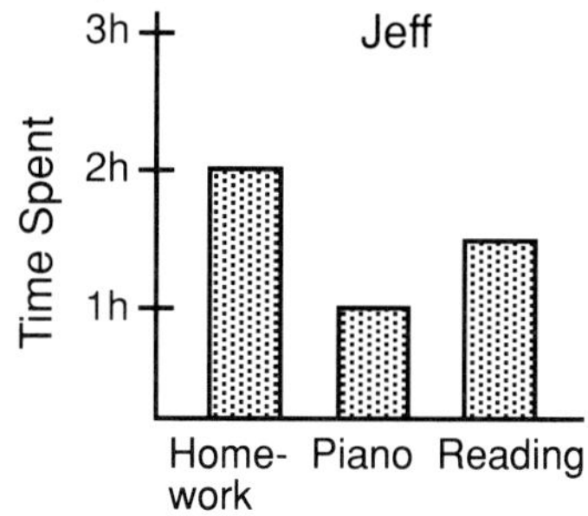

1. How much more time does Linda spend on homework than on piano practice? ________

2. How much more time does Jeff spend on homework than on the piano? ________

3. Do the graphs compare the data fairly? Explain your answer.

____

The table shows how often each score was given on a spelling test. Tell which kind of graph you would make to help you answer each question. Then answer the questions. (Need an example? See pages 400–401.)

Score	100	95	90	85	80	75	70	65 or less
Frequency	4	6	8	16	12	8	4	2

4. Which was the most common score?

____

5. Did more students score 90 or above than scored 75 or less? ______

NAME

Daily Review 116

Tell which type of graph you would make to help you answer the question. Then answer the questions. (Need an example? See pages 404–405.)

Jane compared her scores on spelling tests last month with her scores this month.

March

Week	1	2	3	4
Score	88	80	86	82

April

Week	1	2	3	4
Score	86	90	88	92

1. In which month did her scores change more? ________

Use these two broken-line graphs for Exercise 2. (Need an example? See pages 402–403.)

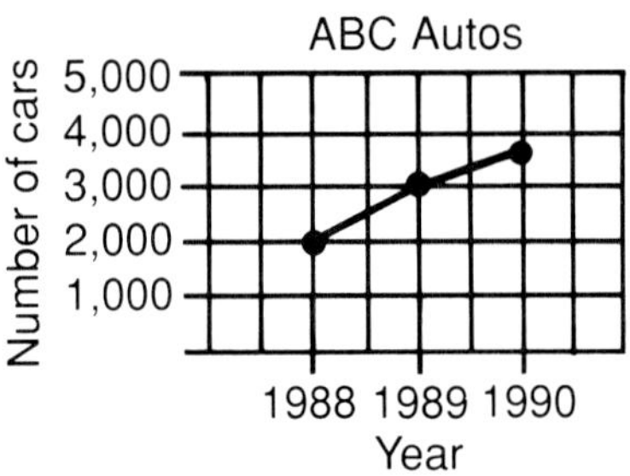

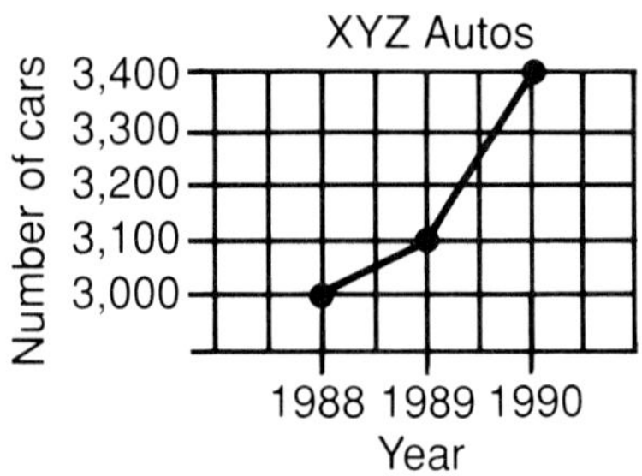

2. Which graph suggests the greater increase in sales? Is the graph misleading? Explain.

____

NAME

Daily Review 117

Draw a probability scale showing the probability of each of the following events occurring. Write the letter corresponding to each event on the scale. (Need an example? See pages 420–423.)

1. The sky (S) will be green tomorrow.
2. You will use a pencil (P) this week.
3. You will pat a dog (D) today.
4. It will rain (R) next Sunday.

Scale

0 ______________________________ 1

For Exercises 5–7 use the table at the right. Draw graphs on a separate sheet of paper. (Need an example? See pages 402–403.)

5. Graph the data so it appears that John had many more points than George.
6. Graph the data so it appears that John had only slightly more points than George.
7. Write a headline for each graph.

Player	Points
John	28
Steve	15
George	25
Max	10

NAME

Daily Review 118

There are 4 pizza parlors in Oakville. A marketing researcher estimated that 500 pizzas are sold in Oakville each Friday. The researcher surveyed 100 residents to see which pizza parlor they preferred. Use the data in the table to predict the number of pizzas that will be sold by each parlor next Friday. (Need an example? See pages 424–425.)

1. Dave's ______________

2. Hilda's ______________

3. Maria's ______________

4. Frank's ______________

Parlor	Number of Votes
Dave's	20
Hilda's	15
Maria's	30
Frank's	35

A box contains green, blue, red, yellow, and white marbles. Joan selects one without looking, records the result and puts back the marble. Use the frequency table to find the probability that the next marble drawn would be the following. For Exercises 5–9, write your answer as a fraction in lowest terms. (Need an example? See pages 420–423.)

5. Green ______________

6. Blue ______________

7. Red ______________

8. Yellow ______________

9. White ______________

Color	Frequency
Green	12
Blue	6
Red	8
Yellow	14
White	16

NAME

Daily Review 119

Jane, Cindy, Jill, Amy, Naomi, Nancy, and Judy each put their names on a slip of paper and placed the slips in a box. One name will be selected. Find each probability. (Need an example? See pages 426–427.)

1. P(Cindy) ________ **2.** P(Amy or Nancy) ________

3. P(Name beginning with J) ________________

4. P(Name beginning with N)________________

An airline questioned 1,000 passengers arriving in Philadelphia about their destinations after they leave the airport. The results are in the table.

Destination	Philadelphia	Pennsylvania (outside Phila.)	New Jersey	Delaware
Passengers	559	201	116	124

Use the data in the table to predict how many of 250,000 passengers who arrive in Philadelphia will go to each destination. (Need an example? See pages 424–425.)

5. Philadelphia ________ **6.** Pennsylvania ________

7. New Jersey ________ **8.** Delaware ________

Use the table at the right to solve Exercise 9. (Need an example? See pages 420–423.)

9. What is the probability that the next number card drawn will be a 3?

____

Card drawn	Frequency
1	10
2	6
3	24
4	10

NAME

Daily Review 120

For Exercises 1–2, tell whether a game with these instructions would be fair. (Need an example? See pages 428–429.)

Each letter of the word *mathematics* is written on an index card. The cards are shuffled three times and placed in a pile face down.

1. You and your partner take turns choosing a card from the pile. You try to guess the letter on the card before turning it over. You get one point for each correct guess.

____

2. The top card is turned over. If it is a vowel, you get one point. If it is a consonant, your partner gets one point.

____

Louis wrote each letter of the word *government* on a different index card. He is about to draw a card. Find each probability. (Need an example? See pages 426–427.)

3. P(consonant) ___ **4.** P(vowel) ___ **5.** P(O or N) ___

Out of the Riverview High School student population, 100 were asked to choose an elective for next year. Use the data in the table to predict how many of all 2,000 students would choose each course next year. (Need an example? See pages 424–425.)

Course	Number
Astronomy	20
Writing	45
Russian	10
Computer	25

6. Astronomy ______________ **7.** Russian ______________

8. Writing ______________ **9.** Computer ______________

NAME

Daily Review 121

Students who attend Sports' Day can choose from 5 water sports, 8 track-and-field events, 4 team sports, and 6 gymnastic events. With the following options, how many choices are there? (Need an example? See pages 430–431.)

1. Water sport, team sport, and gymnastic event ______
2. Track-and-field event and gymnastic event ______
3. Water sport, track-and-field event, gymnastic event ______

For Exercises 4–5 tell whether a game with these instructions would be fair. (Need an example? See pages 428–429.)

Numbers from 1 to 9 are written one to an index card. A card is chosen without looking.

4. If the card shows an even number, you get one point. If the card shows an odd number, your partner gets one point.

   ______

5. If the card shows a number less than 4, you get a point. If it shows a number greater than 6, your partner gets a point. No points are given for other numbers.

   ______

A number is selected from the set of nonzero multiples of 5 less than 100. Find each probability. Need an example? See pages 426–427.)

6. P(even number) ______
7. P(multiple of 4) ______

NAME

Daily Review 122

The table at the right shows the probability that a type of seed from Dot Seed Co. will germinate. What is the probability, to the nearest hundredth, that two seeds of the given type will germinate? (Need an example? See pages 434–435.)

P(germination)	
Zucchini	0.9
Tomato	0.87
Marigold	0.91
Impatiens	0.76

1. Zucchini ____________

2. Tomato ____________

3. Marigold ____________

4. Impatiens ____________

A store contains 43 styles of shirts, 15 styles of slacks, 21 styles of skirts and 30 styles of belts. With the following options, how many choices are there? (Need an example? See pages 430–431.)

5. Shirt and slacks ____________

6. Shirt, slacks, and belt ____________

7. Belt and skirt ____________

8. Shirt, skirt, and belt ____________

A spinner is divided into 8 equal parts. The parts are labeled 12, 8, 45, 81, 77, W, H, and A. Find each probability. (Need an example? See pages 426–427.)

9. P(multiple of 3) ______ **10.** P(letter) ______

11. P(number > 50) ______ **12.** P(consonant) ______

NAME

Daily Review 123

Suppose that the numbers from 1 to 25 are written on cards, one to a card, and the 25 cards are shuffled. If you draw 2 cards and do not replace them, find the theoretical probability of drawing the following cards. (Need an example? See pages 436–437.)

1. An even, then an odd ____________

2. A multiple of 4, then 11 ____________

3. A multiple of 10, then a multiple of 2 ____________

4. Two numbers less than 9 ____________

A bag contains 6 green cards, 5 yellow cards, and 4 black cards. A card is drawn, its color recorded, and the card is replaced. Find each probability. (Need an example? See pages 434–435.)

5. P(Y, Y) ____________

6. P(G, B) ____________

7. P(B, Y) ____________

8. P(G, G) ____________

9. P(Y, G) ____________

10. P(Y or G, B) ____________

Solve the problem. (Need an example? See pages 430–431.)

11. An answering service gives each of its customers a code that is a four-digit number. The first digit must be 1, 2, or 3 and the last digit cannot be 0. How many different codes can the service give out? ____________

NAME

Daily Review 124

Each letter of the word *needed* is written on an index card. The cards are shuffled, one card is drawn, recorded, replaced, and another card is drawn. Draw a probability tree diagram and use it to find each probability. (Need an example? See pages 438–439.)

1. P(D, D) __________ **2.** P(E, N) __________

3. P(D, E) __________ **4.** P(N, N) __________

5. P(E, E) __________ **6.** P(N, D) __________

Suppose that a box contains 9 silver paper clips (S), 8 bronze clips (B), and 5 red clips (R). If you draw 3 clips and do not replace them, find the theoretical probability of drawing the following sequences of clips. (Need an example? See pages 436–437.)

7. S, B, R __________ **8.** S, S, S __________

9. B, R, B __________ **10.** R, R, S __________

11. B, B, B __________ **12.** B, R, R __________

Scouts brought a box to camp that has 15 cans of apple juice, 25 cans of orange juice, 10 cans of pineapple juice, and 5 cans of grape juice. A can is chosen, the flavor noted, and then replaced in the box. Find each probability. (Need an example? See pages 434–435.)

13. P(A, O) __________ **14.** P(P, G) __________

15. P(A, G) __________ **16.** P(A, A) __________

17. P(O, G) __________ **18.** P(G, G) __________

NAME

Daily Review 125

Use the random number table on page 440 to simulate the following experiments.
Make a table on another sheet of paper to show the results of each simulation. (Need an example? See pages 440–441.)

1. A number cube has the digits 1–6 on each side. It is tossed 60 times. How often does each number come up?

____

2. A spinner has 10 equal regions labeled A–J. It is spun 200 times. How often does each letter come up?

____

Alfred put a red mark on 3 cards, a blue mark on 5 cards, and a yellow mark on 2 cards. He will place the cards in a bag, draw one, record the color mark, replace it, and draw another card. Draw a probability tree diagram and use it to find each probability. (Need an example? See pages 438–439.)

3. P(R, R) _____ 4. P(R, B) _____ 5. P(B, B) _____

6. P(Y, B) _____ 7. P(Y, Y) _____ 8. P(Y, R) _____

The letters in the word *eligible* are written one to a card. Suppose that the cards are shuffled. If you draw three cards and do not replace them, find the theoretical probability of drawing the following. (Need an example? See pages 436–437.)

9. E, B, L _____ 10. L, L, E _____ 11. G, I, I _____

12. L, E, E _____ 13. B, G, L _____ 14. I, G, E _____

15. L, I, E _____ 16. G, E, G _____ 17. B, I, G _____

NAME ____________________

Daily Review 126

Express each quantity as an integer. (Need an example? See pages 452–453.)

1. 1,000 ft above sea level ________
2. An increase of 24 ________
3. A profit of \$95 ________
4. A loss of \$125 ________
5. A penalty of 10 yd ________
6. 40 degrees above zero ________

Use the random number table on page 440 to simulate the following experiments. Make a table on another sheet of paper to show the results of each simulation. (Need an example? See pages 440–441.)

7. A coin is tossed 50 times.

____

8. A spinner has 4 equal regions numbered 1–4. It is spun 100 times. How often does each number come up?

____

A set of cards contains 9 with a drawing of a circle, 3 with a drawing of a triangle and 4 with a drawing of a square. Two cards are drawn and their shapes recorded without replacement. Draw a probability tree diagram and use it to find each probability. (Need an example? See pages 438–439.)

9. P(C, T) ______
10. P(C, C) ______
11. P(T, S) ______
12. P(C, S) ______
13. P(S, S) ______
14. P(T, T) ______

NAME

Daily Review 127

Add. Use a number line if necessary. (Need an example? See pages 454–455.)

1. $6 + 18 =$ ____________ **2.** $-5 + (-7) =$ ____________

3. $-8 + 4 =$ ____________ **4.** $-9 + (-1) =$ ____________

5. $-6 + 4 =$ ____________ **6.** $-7 + (-3) =$ ____________

7. $-12 + 11 + (-16) =$ ____________

8. $-3 + (-4) + 5 =$ ____________

Write the absolute value.
(Need an example? See pages 452–453.)

9. -4 ____ **10.** $+53$ ____ **11.** $+27$ ____ **12.** -60 ____

Use the random number table on page 440 to simulate the following experiments. Make a table on another sheet of paper to show the results of each simulation. (Need an example? See pages 440–441.)

13. A number cube has the digits 1–6. It is tossed 90 times. How often does each number come up?

____

14. A spinner has 8 equal regions numbered 1–8. It is spun 40 times. How often does each number come up?

____

15. A coin is tossed 20 times.

____

NAME

Daily Review 128

First write each subtraction as an addition. Then find the answer. (Need an example? See pages 456–457.)

1. $-3 - (-3)$ ______ **2.** $81 - (-11)$ ______

3. $-35 - (-45)$ ______ **4.** $-53 - 12$ ______

5. $41 - (-7)$ ______ **6.** $19 - 15$ ______

7. $0 - 67$ ______ **8.** $0 - (-92)$ ______

Add. Use a number line if necessary. (Need an example? See pages 454–455.)

9. $5 + 17 =$ ______ **10.** $-12 + (-4) =$ ______

11. $-7 + (-9) =$ ______ **12.** $-9 + 5 =$ ______

13. $-6 + (-8) =$ ______ **14.** $14 + (-17) =$ ______

15. $-5 + (-5) + 2 =$ ______ **16.** $8 + (-4) + 7 =$ ______

Express each quantity as an integer. (Need an example? See pages 452–453.)

17. 15 strokes over par ______ **18.** Forward 12 ______

19. 2 strokes under par ______ **20.** Backwards 3 ______

21. 150 ft above sea level ______ **22.** 12 degrees below 0 ______

23. A profit of $2,000 ______ **24.** A decrease of 50 ______

25. A loss of $350 ______ **26.** A gain of 34 yd ______

NAME

Daily Review 129

Multiply.
(Need an example? See pages 458–459.)

1. $15 \times 60 =$ ________ **2.** $-21 \times 7 =$ ________

3. $18 \times (-14) =$ ________ **4.** $-9 \times 50 =$ ________

5. $-23 \times 8 =$ ________ **6.** $(-25)(-4) =$ ________

7. $(4)(-8) =$ ________ **8.** $(12)(-10) =$ ________

9. $(-11)(-15) =$ ________ **10.** $(-90)(30) =$ ________

Subtract. (Need an example? See pages 456–457.)

11. $-8 - (-8) =$ ________ **12.** $32 - (-17) =$ ________

13. $86 - (-40) =$ ________ **14.** $0 - (-39) =$ ________

15. $-61 - 62 =$ ________ **16.** $-34 - 43 =$ ________

17. $18 - 23 =$ ________ **18.** $-29 - (-44) =$ ________

Add. Use a number line if necessary. (Need an example? See pages 454–455.)

19. $9 + 32 =$ ________ **20.** $(-20) + 7 =$ ________

21. $-3 + (-9) =$ ________ **22.** $-9 + 4 =$ ________

23. $-7 + (-11) =$ ________ **24.** $27 + (-31) =$ ________

25. $-18 + 7 =$ ________ **26.** $-78 + (-15) =$ ________

Which is greater? (Need an example? See pages 452–453.)

27. -33 or 11 ________ **28.** -140 or -14 ________

NAME ____________________

Daily Review 130

Divide. **Remember** to check the sign in the answer. (Need an example? See pages 460–461.)

1. $-30 \div (-6) =$ ________ **2.** $-72 \div 6 =$ ________ **3.** $-32 \div (-4) =$ ________

4. $49 \div (-7) =$ ________ **5.** $-40 \div 8 =$ ________ **6.** $-64 \div (-8) =$ ________

7. $\frac{-35}{-7}$ ________ **8.** $\frac{45}{-9}$ ________ **9.** $\frac{-135}{-3}$ ________

10. $\frac{-124}{4}$ ________ **11.** $\frac{-100}{5}$ ________ **12.** $\frac{-555}{-5}$ ________

Multiply. (Need an example? See pages 458–459.)

13. $(100)(17) =$ ________ **14.** $-5 \times 30 =$ ________

15. $6 \times (-60) =$ ________ **16.** $(-29)(-7) =$ ________

17. $8 \times (-35) =$ ________ **18.** $-4 \times 19 =$ ________

19. $(12)(-14) =$ ________ **20.** $(-36)(-21) =$ ________

21. $(13)(-18) =$ ________

Subtract. **Remember** to write each subtraction as an addition problem first. (Need an example? See pages 456–457.)

22. $-4 - (-4) =$ ________ **23.** $32 - (-15) =$ ________

24. $19 - (-7) =$ ________ **25.** $91 - (-91) =$ ________

26. $-14 - (-21) =$ ________ **27.** $0 - 73 =$ ________

28. $-41 - 41 =$ ________ **29.** $0 - (-86) =$ ________

NAME ____________

Daily Review 131

Find the time and height of each high tide on the given date in the given city. Use the tables on page 464 of your textbook. (Need an example? See pages 464–465.)

1. Cape May, NJ, June 5, evening tide

____

2. Havre de Grace, MD, June 1, morning tide

____

3. Annapolis, MD, June 3, evening tide

____

Divide. **Remember** to check the sign in the answer. (Need an example? See pages 460–461.)

4. $-55 \div (-5) =$ ______ **5.** $28 \div (-7) =$ ______ **6.** $-56 \div (-8) =$ ______

7. $\frac{-32}{4}$ ______ **8.** $\frac{-18}{3}$ ______ **9.** $\frac{-21}{7}$ ______

10. $\frac{-36}{-9}$ ______ **11.** $\frac{-48}{12}$ ______ **12.** $\frac{-350}{-7}$ ______

Multiply. (Need an example? See pages 458–459.)

13. $12 \times (-16)$ ______ **14.** $(-40) \times 5$ ______

15. $-8 \times (-19)$ ______ **16.** $(70)(-3)$ ______

17. $(-18)(5)$ ______ **18.** $(-27)(-4)$ ______

19. $(-2)(-7)(-4)$ ______ **20.** $(8)(9)(-2)$ ______

21. $(-5)(-5)(-5)$ ______ **22.** $(-6)(-7)(4)$ ______

NAME

Daily Review 132

Compare. Use <, >, or =.
(Need an example? See pages 466–467.)

1. $\frac{-1}{5}$ ◯ $\frac{-1}{9}$ **2.** 0 ◯ $\frac{-1}{10}$ **3.** $\frac{7}{5}$ ◯ −3

4. $\frac{-1}{7}$ ◯ −1.7 **5.** $\frac{-3}{5}$ ◯ $\frac{-6}{10}$ **6.** −6 ◯ $-8\frac{3}{9}$

Find the time and height of each high tide on the given date in the given city. Use the table on page 464 of your textbook. (Need an example? See pages 464–465.)

7. Atlantic City, NJ, June 4, morning tide

____

8. Cape May, NJ, June 1, evening tide

____

9. Havre de Grace, MD, June 5, morning tide

____

Divide. **Remember** to check the sign in the answer. (Need an example? See pages 460–461.)

10. $\frac{20}{-5}$ ________ **11.** $\frac{-72}{-8}$ ________ **12.** $\frac{27}{-9}$ ________

13. $\frac{-33}{11}$ ________ **14.** $\frac{2,500}{-50}$ ________ **15.** $\frac{-20,000}{-4}$ ________

16. $\frac{450}{-9}$ ________ **17.** $\frac{-7,650}{-3}$ ________ **18.** $\frac{56,000}{-8}$ ________

Multiply. (Need an example? See pages 458–459.)

19. $13 \times (-7) =$ ________ **20.** $(-22)(-24) =$ ________

21. $(4) \times (61) =$ ________ **22.** $-15 \times 20 =$ ________

23. $-248 \times 0 =$ ________ **24.** $-35 \times -6 =$ ________

NAME

Daily Review 133

Between which two consecutive integers is each of the following? (Need an example? See pages 468–469.)

1. $\sqrt{32}$ ______ **2.** $\sqrt{61}$ ______ **3.** $\sqrt{29}$ ______

4. $\sqrt{19}$ ______ **5.** $\sqrt{8}$ ______ **6.** $\sqrt{42}$ ______

7. $\sqrt{6}$ ______ **8.** $\sqrt{53}$ ______ **9.** $\sqrt{76}$ ______

10. $\sqrt{3}$ ______ **11.** $\sqrt{70}$ ______ **12.** $\sqrt{12}$ ______

13. $\sqrt{105}$ ______ **14.** $\sqrt{98}$ ______ **15.** $\sqrt{84}$ ______

Compute. (Need an example? See pages 466–467.)

16. $-\frac{1}{4} + \left(-3\frac{1}{6}\right) =$ ______

17. $(-5)\left(\frac{1}{3}\right) =$ ______

18. $-6.93 + 7 =$ ______

19. $-2\frac{1}{8} - 1\frac{3}{4} =$ ______

20. $9\frac{2}{3} - \frac{1}{3} =$ ______

21. $4\frac{1}{2} - (-6) =$ ______

22. $\frac{6}{11} \times \left(-1\frac{7}{15}\right) =$ ______

23. $3\frac{3}{4} \div 1\frac{1}{2} =$ ______

24. $6.4 - (-0.8) =$ ______

25. $(-5)\left(\frac{-3}{4}\right) =$ ______

26. $9.1(-3) =$ ______

27. $\frac{-1}{4} + \frac{1}{8} =$ ______

Find the time and height of each high tide on the given date in the given city. Use the tables on page 464 of your textbook. (Need an example? See pages 464–465.)

28. Bar Harbor, ME, June 1, morning tide

______

29. Coney Island, NY, June 4, evening tide

______

NAME ____________________

Daily Review 134

In Exercise 1–8, a, b, and c are lengths of sides in a right triangle. Find the missing length. Round decimal answers to the nearest tenth. (Need an example? See pages 470–473.)

1. $a = 4$, $b = 5$, $c =$ _____ **2.** $a = 6$, $c = 10$, $b =$ _____

3. $b = 4$, $c = 9$, $a =$ _____ **4.** $a = 7$, $b = 5$, $c =$ _____

5. $a = 12$, $c = 20$, $b =$ ____________________

6. $b = 120$, $c = 150$, $a =$ ____________________

7. $a = 10$, $b = 20$, $c =$ ____________________

8. $a = 25$, $b = 42$, $c =$ ____________________

Estimate the positive square root of each number to the nearest integer. (Need an example? See pages 468–469.)

9. 10 _____ **10.** 86 _____ **11.** 135 _____ **12.** 73 _____

13. 31 _____ **14.** 97 _____ **15.** 140 _____ **16.** 15 _____

17. 45 _____ **18.** 61 _____ **19.** 118 _____ **20.** 5 _____

Show that each number is a rational number by writing it as a ratio of two integers. (Need an example? See pages 466–467.)

21. 8 __________ **22.** -5.2 __________ **23.** $-9\frac{1}{6}$ __________

24. -3 __________ **25.** 6.75 __________ **26.** $5\frac{11}{12}$ __________

27. -7.1 __________ **28.** 2.07 __________ **29.** -8.301 __________

NAME

Daily Review 135

Solve each problem. Use a calculator if necessary. Give your answers to the nearest tenth. (Need an example? See pages 474–475.)

1. The park's map shows the distance from the fountain to the monument, and from the gate to the monument. Find the distance from the gate to the fountain.

____

2. A 30-ft ladder is 15 feet from the base of a building. How high up the building can the ladder reach?

____

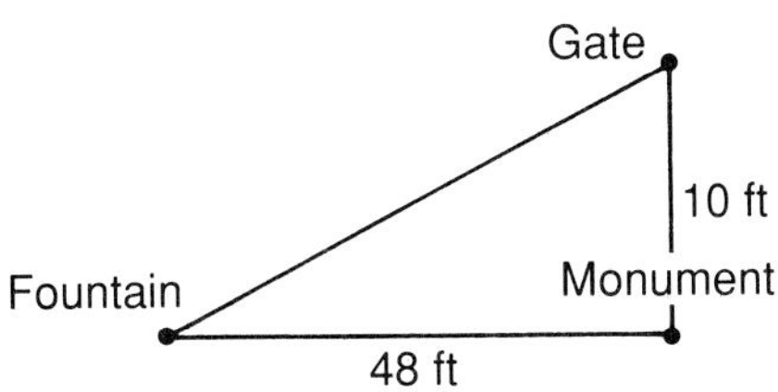

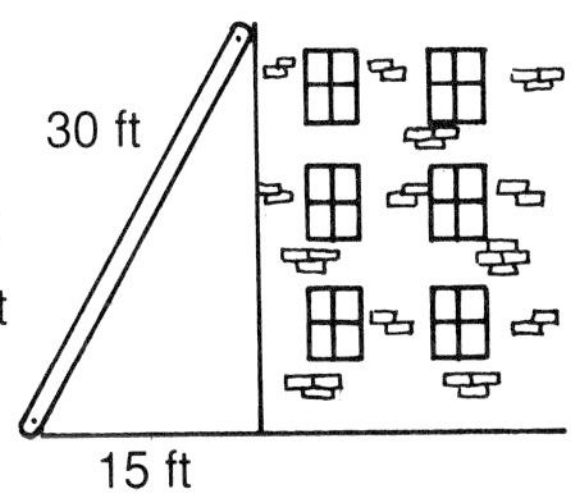

In Exercises 3–6, *a, b,* and *c* are lengths of sides in a right triangle. Find the missing length. Round decimal answers to the nearest tenth. (Need an example? See pages 470–473.)

3. $a = 21$, $c = 35$, $b =$ ___

4. $a = 7$, $b = 8$, $c =$ ______

5. $b = 11$, $c = 30$, $a =$ ___

6. $a = 3$, $c = 10$, $b =$ ____

Use your calculator or the table on page 468 to help find each square root. Round each answer to the nearest tenth. (Need an example? See pages 468–469.)

7. $\sqrt{50}$ __________

8. $\sqrt{29}$ __________

9. $\sqrt{(14)^2}$ ________

NAME ____________________

Daily Review 136

Write each number with an exponent. (Need an example? See pages 476–477.)

1. 10,000 ______ **2.** 1,000,000 ______ **3.** $\frac{1}{1,000}$ ______

4. 0.00001 ______ **5.** 0.0000001 ______ **6.** 10,000,000 ______

Solve each problem. Use a calculator if you wish. Round answers to the nearest tenth. (Need an example? See pages 474–475.)

7. A guy wire is attached to a flag pole at a point 20 ft from the base of the pole. The pole is 35 ft high. How long is the guy wire?

____

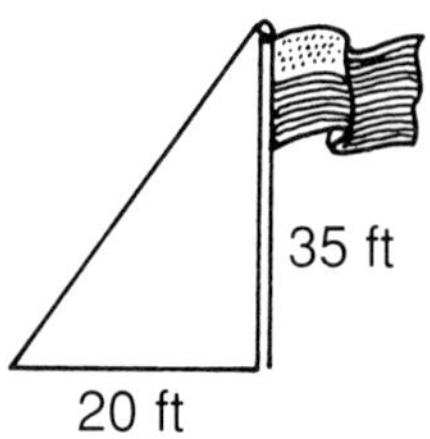

8. Jill wants to reach the second floor window with a ladder that measures 25 ft. How far from the base of the house should she place the base of the ladder?

____

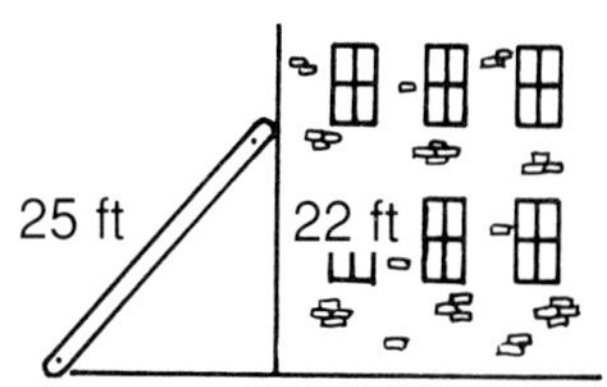

In Exercises 9–12, a, b, and c are lengths of sides in a right triangle. Find the missing length. Round decimal answers to the nearest tenth. (Need an example? See pages 470–473.)

9. $a = 15$, $b = 20$, $c =$ ___ **10.** $a = 7$, $c = 31$, $b =$ ___

11. $a = 10$, $b = 15$, $c =$ ___ **12.** $b = 33$, $c = 55$, $a =$ ___

NAME

Daily Review 137

Write each number in scientific notation.
(Need an example? See pages 478–479.)

1. 406,000 ______ **2.** 3,291,000 ______

3. 56.2 ______ **4.** 420.18 ______

Write in expanded form. (Need an example? See pages 476–477.)

5. 2,473,112 ______

6. 27.108 ______

7. 0.0921 ______

Solve. Use a calculator if you wish. Round your answer to the nearest tenth. (Need an example? See pages 474–475.)

8. The map shows a park. The path from the entrance to the playground measures 16 m. The path from the playground to the rink measures 24 m. How long is the path from the entrance to the rink?

______

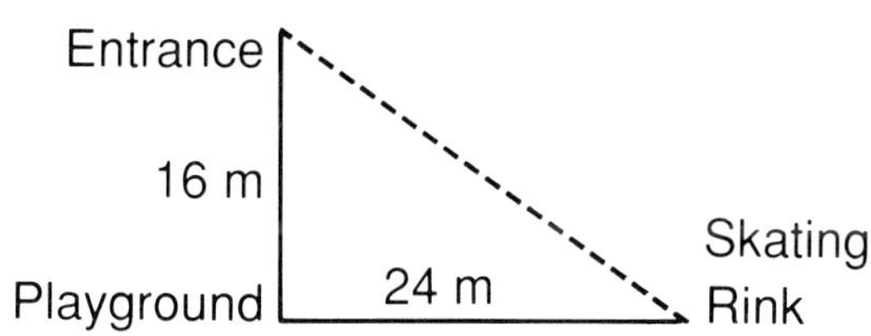

NAME

Daily Review 138

Compute each answer. State which property or properties you used. (Need an example? See pages 490–493.)

1. $5\left(\frac{3}{10} + \frac{2}{5}\right)$ ____________

2. $3.46 \times 6 + 3.46 \times 4$ ____________

3. $(-7.2 \times 1) + (4 \times 0)$ ____________

4. $25 \times 3.12 \times 4$ ____________

Write each number in standard form. (Need an example? See pages 478–479.)

5. 2.8×10^3 ____________

6. 4.07×10^2 ____________

7. 6.88×10^5 ____________

8. 3.214×10^7 ____________

9. 4.6×10^0 ____________

10. 7.73×10^8 ____________

Write in expanded form. (Need an example? See pages 476–477.)

11. 5,432,897 ________________________

12. 16,097,432 ________________________

13. 2.096 ________________________

14. 0.00987 ________________________

15. 3.8009 ________________________

NAME ____________________

Write an expression for each phrase. (Need an example? See pages 494–495.)

1. 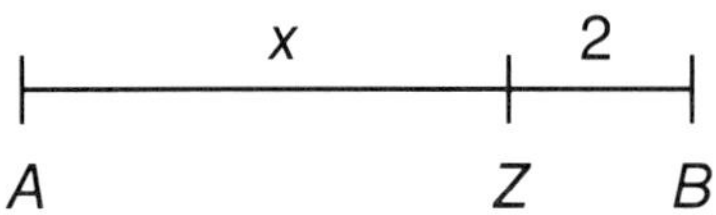

The length of AB ______

2. 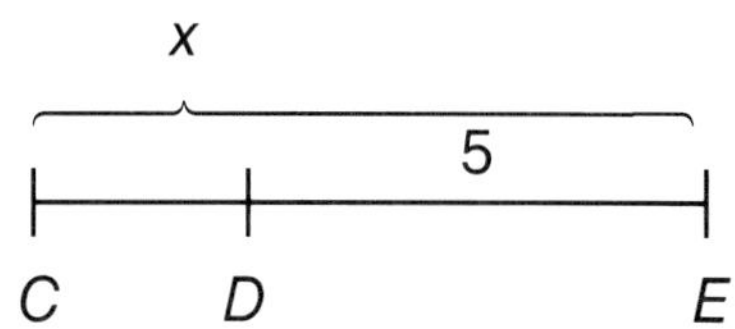

The length of CD ______

3. The opposite of the sum of x and y ____________

4. \$15 more than the price, p ____________

Find the value of the variable that makes the statement true. Name the property or properties you used. (Need an example? See pages 490–493.)

5. $0.21y = 0.21$

____

6. $[3.4 \times (-2.1)] \times 4.9 = 3.4 \times [-2.1 \times t]$

____

7. $(4.3)(-5) = -5h$

____

8. $5(-2 + 11) = -10 + s$

____

Solve each problem. (Need an example? See pages 478–479.)

9. The last of the dinosaurs disappeared about 65 million years ago. Write that number in scientific notation.

____

10. The speed of light is about 1.86×10^5 miles per second. Write this number in standard form.

____

  Use before pages 496–497.

NAME

Daily Review 140

Solve each equation. Check your answer.
(Need an example? See pages 496–497.)

1. $a - 7.4 = -20.2$ ______

2. $b - 1.2 = 8.8$ ______

3. $c + \frac{2}{5} = \frac{-1}{10}$ ______

4. $-15.1 + d = 15.1$ ______

5. $\frac{3}{8} + e = \frac{1}{2}$ ______

6. $-12.05 = 7.32 + f$ ______

Use the figure at the right to write an expression. (Need an example? See pages 494–495.)

7. The distance from *A* to *C* going through *B* ______

8. The distance saved by going directly from *A* to *C* rather than through *B* ______

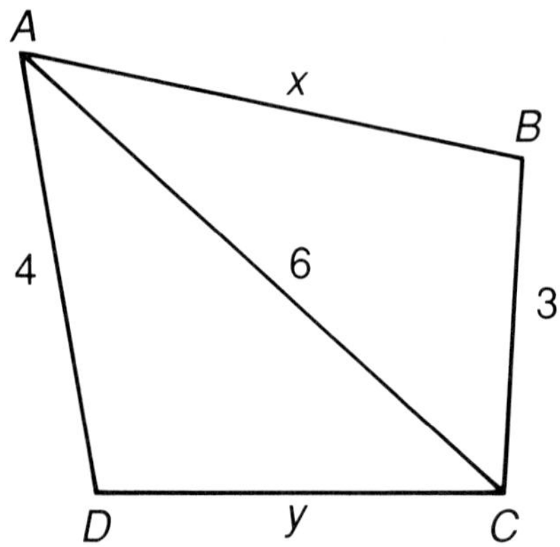

9. The distance from *C* to *D* ______

10. The difference between the distance from *A* to *C* going through *B* and the distance from *C* to *D* ______

NAME

Daily Review 141

Evaluate each expression when $x = 0.6$.
(Need an example? See pages 498–499.)

1. $22 - x$ ______ **2.** $\frac{5x}{-4 - 1}$ ______ **3.** $\frac{10x - 3}{3}$ ______

4. $x^2 + 7$ ______ **5.** $-3x^2$ ______ **6.** $(2x)^2 + 5x$ ______

Solve each equation. Check your answer. (Need an example? See pages 496–497.)

7. $\frac{6}{10} + k = \frac{4}{5}$ ______ **8.** $m - 2.7 = 5.3$ ______

9. $-54.2 + n = 54.2$ ______ **10.** $6.18 + p = -8.38$ ______

11. $4\frac{2}{5} + q = -2\frac{1}{10}$ ______ **12.** $5.1 - r = 8.6$ ______

Evaluate each expression when $x = -2$ and $y = 0.1$. (Need an example? See pages 494–495.)

13. $12 - y + 3x$ ______ **14.** $\frac{3y - 2}{x}$ ______ **15.** $4y + (-x)$ ______

16. $\frac{50y}{-10x}$ ______ **17.** $x^2 - 300y$ ______ **18.** $y^3 - x^3$ ______

19. $-6x + 20y$ ______ **20.** $x^2 - y^2$ ______ **21.** $\frac{x^2 + y}{x}$ ______

22. $x^3 + y$ ______ **23.** $\left(\frac{x}{y}\right)^2$ ______ **24.** $10x + 10y$ ______

  Use before pages 502–503.

NAME

Daily Review 142

Solve each equation. (Need an example? See pages 502–503.)

1. $-5a = 95$
2. $\left(\frac{-1}{4}\right)b = \frac{1}{3}$
3. $-0.02c = 6.2$
4. $\left(\frac{-3}{5}\right)d = \frac{-5}{6}$
5. $\frac{4}{9} = \left(\frac{-2}{9}\right)e$
6. $\frac{d}{-2.4} = 3$

Evaluate each expression when $b = \frac{2}{3}$ and $c = -3$. (Need an example? See pages 498–499.)

7. $\frac{6b}{7} + \frac{c}{7}$
8. $-b(c + (-9))$
9. $\frac{c(b + \frac{1}{6})}{2}$
10. $6b - \left(\frac{2}{3}\right)c$
11. $\left(\frac{1}{4}\right)b + \left(\frac{-2}{9}\right)c$
12. $\frac{-b + \frac{5}{6}}{c}$

Solve each problem. **Remember** that Sales − Expenses = Profit, $S - E = P$. (Need an example? See pages 496–497.)

13. Mrs. Lee's florist shop has monthly expenses of $3,500. Her profit was $6,000. What were her sales?

14. The tennis shop earned a profit of $5,600 last month. Its sales were $8,100. What were its expenses?

NAME

Daily Review 143

Write an equation for each problem. Then solve it. (Need an example? See pages 504–505.)

1. A plane travels with the wind at 560 miles per hour. The speed of the plane with no wind is 525 miles per hour. What is the wind speed?

______

2. Jeremy is driving to visit his grandparents 2,500 miles away. He plans to drive 400 miles per day. How many days will the trip take?

______

Solve each equation. (Need an example? See pages 502–503.)

3. $-5g = 32$ ______

4. $1.5h = -600$ ______

5. $2j = -3.47$ ______

6. $\frac{-k}{10} = 4.32$ ______

7. $\frac{-9m}{10} = \frac{-1}{2}$ ______

8. $4n = -3\frac{2}{3}$ ______

Evaluate each expression when $x = -0.5$ and $y = 4$. (Need an example? See pages 498–499.)

9. $-12x + y^2$ ______

10. $3x + (-7)y$ ______

11. $\frac{8x - 3y}{-2}$ ______

12. $\frac{-2y}{6 + (-4)}$ ______

13. $4x - 6y$ ______

14. $y^2 - 5x$ ______

15. $y - 16x^2$ ______

16. $\frac{3y}{2x - 1}$ ______

17. $x^2 + y^2$ ______

  Use before pages 506–507.

NAME

Daily Review 144

Solve each equation. (Need an example? See pages 506–507.)

1. $2a + 11 = 39$ ______

2. $5b - 3 = 42$ ______

3. $\frac{c}{8} + 12 = 20$ ______

4. $\frac{d}{6} - 10 = 4$ ______

5. $4e - 31 = 1$ ______

6. $3f + 24 = 36$ ______

Solve each problem. (Need an example? See pages 504–505.)

7. A plane flew a distance of 840 miles in $2\frac{1}{2}$ hours. What was the speed of the plane?

______

8. The distance between Mary's house and Juanita's house is 3.2 miles. The distance between Mary's house and Frieda's house is $\frac{1}{4}$ of the distance to Juanita's. What is the distance from Mary's to Frieda's?

______

Solve each equation. (Need an example? See pages 502–503.)

9. $-7w = 6.3$ ______

10. $4x = -1.28$ ______

11. $-3.9 = 7.8y$ ______

12. $\frac{x}{-12} = -9$ ______

13. $-7 = \frac{-287}{r}$ ______

14. $-4p = 2\frac{2}{3}$ ______

NAME

Daily Review 145

A veterinarian uses the following formula to determine how much of a certain medicine to prescribe for a dog: $\frac{10 \times p}{3}$. This tells her the number of milligrams per dose based on the dog's weight in pounds (p). Use this information to solve problems 1–2. (Need an example? See pages 508–509.)

1. How much medicine should be put in each dose for a dog that weighs 75 lb?

____

2. A dog has been given a dose of 150 milligrams. How much does the dog weigh?

____

Solve each equation. (Need an example? See pages 506–507.)

3. $2a - 11 = 9$ ____________

4. $4b + 3 = 27$ ____________

5. $3c - 12 = 48$ ____________

6. $\frac{d}{5} - 6 = 2$ ____________

7. $\frac{e}{6} + 3 = 13$ ____________

8. $15 = 4 + \frac{f}{2}$ ____________

Solve. (Need an example? See pages 504–505.)

9. An airplane flies from Seaside Airport to Valleyshow Airfield in $6\frac{1}{2}$ hours. The speed of the plane is 525 miles per hour. How far is Seaside from Valleyshow? ____________

NAME

Daily Review 146

Describe in words the sets of numbers graphed below.
(Need an example? See pages 520–521.)

1.

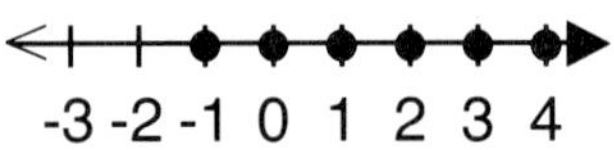

____

2.

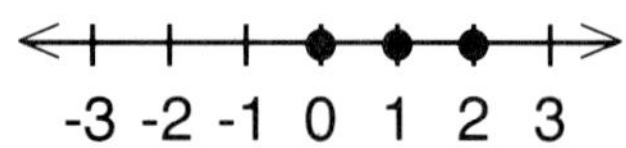

____

3.

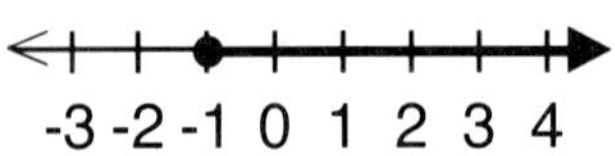

____

4.

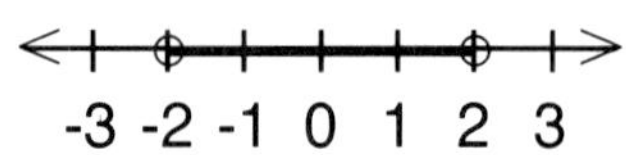

____

Steve is paid \$8.00 per hour for a 40-hour week. For any time he works over 40 hours, he is paid one-and-a-half times his regular rate. He uses the formula $320 + 12(h) = w$, where h is the number of hours over 40 that he has worked in a given week and w is his wage for the week.
(Need an example? See pages 508–509.)

5. How much are his wages if he works 46 hours one week?

____

6. One week he earned \$380. How many hours had he worked?

____

Solve each equation. (Need an example? See pages 506–507.)

7. $5a - 7 = 28$ ________

8. $6b - 4 = 26$ ________

NAME

Daily Review 147

Name the point located by each ordered pair. (Need an example? See pages 522–523.)

1. (−1, 3) ________

2. (2, 2) ________

3. (0, −4) ________

4. (−3, −2) ________

5. (3, −1) ________

6. (−4, 0) ________

7. (−2, −3) ________

8. (1, 2) ________

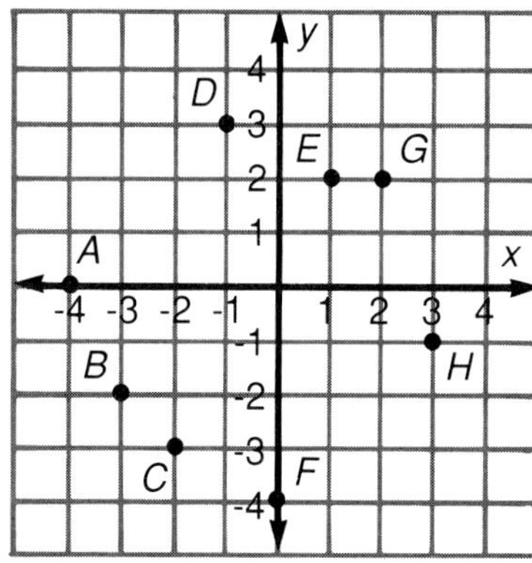

Graph the following sets of numbers on a separate sheet of paper. (Need an example? See pages 520–521.)

9. Whole numbers greater than 3

10. Integers between 3 and −3

11. All numbers less than 5

12. Integers greater than −5

A mason uses the formula $6.5s = b$ to estimate the number of bricks (b) needed given the number of square feet (s) that must be built. Use this formula to solve these problems. (Need an example? See pages 508–509.)

13. How many bricks are needed for a wall that will measure 75 square feet?

____

14. How many square feet can be covered using 1,300 bricks?

____

NAME

Daily Review 148

Copy $\triangle EFG$ on grid paper. On the same set of axes, graph its image under the translations described below. Give the coordinates of the vertices. (Need an example? See pages 524–527.)

1. A vertical translation of -3 units

____

2. A horizontal translation of 2 units

____

3. A horizontal translation of -1 units and a vertical translation of 2 units

____

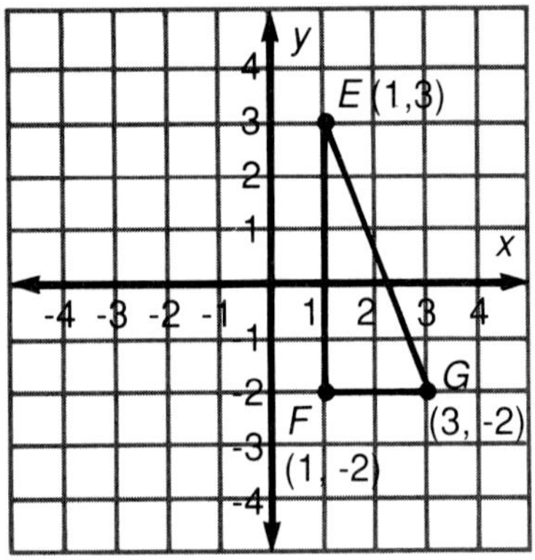

Give the coordinates of the following points. (Need an example? See pages 522–523.)

4. *A* ______________________

5. *B* ______________________

6. *C* ______________________

7. *D* ______________________

8. *E* ______________________

9. *F* ______________________

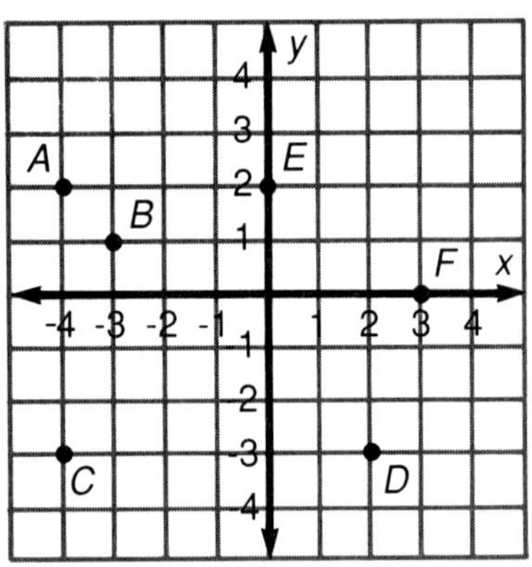

NAME

Daily Review 149

Refer to the figure at the right to answer the following questions. (Need an example? See pages 528–531.)

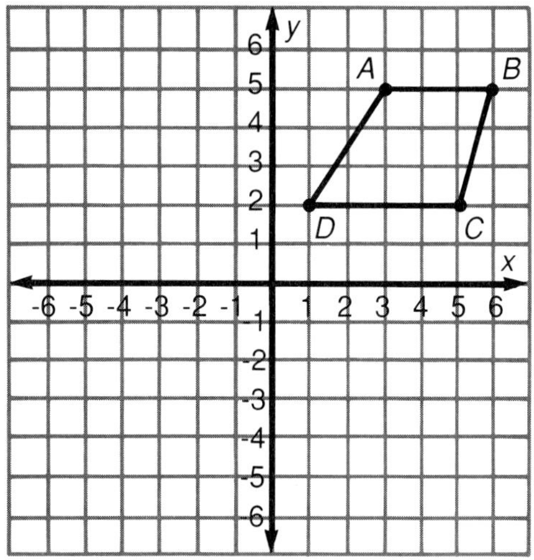

1. If you reflect *ABCD* over the *x*-axis, what are the coordinates of the vertices of the image?

____

2. If you reflect *ABCD* over the *y*-axis, what are the coordinates of the vertices of the image?

____

A rectangle has coordinates of $A(-1, 3)$, $B(2, 3)$, $C(2, -4)$, and $D(-1, -4)$. Find the coordinates of the vertices for each translation below. (Need an example? See pages 524–527.)

3. 3 units to the left

____

4. 2 units to the right

____

5. 1 unit down

____

6. 1 unit up and 3 units to the right

____

NAME

Daily Review 150

Copy figure *ABCD* onto your graph paper. Give the coordinates of the vertices of its image under each of the following rotations. (Need an example? See pages 532–535.)

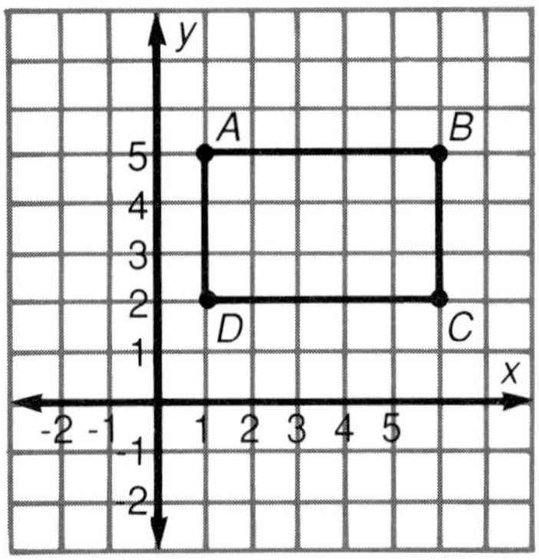

1. 90° counterclockwise rotation

____

2. 180° clockwise rotation

____

Refer to the figure to answer the question. (Need an example? See pages 528–531.)

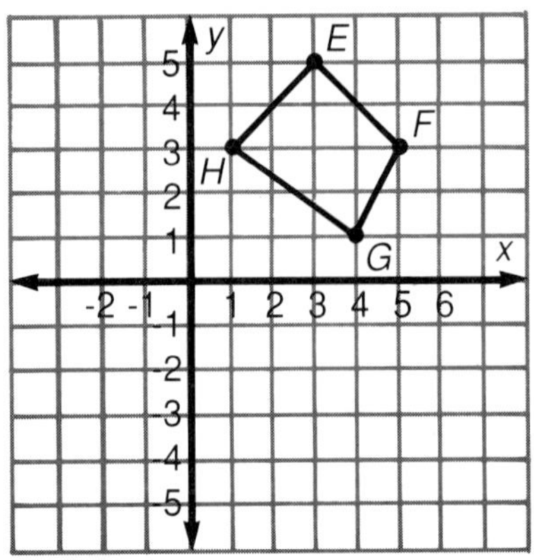

3. If you reflect *EFGH* over the *x*-axis, what are the coordinates of the vertices of the image?

____

Use graph paper to solve the following problem. (Need an example? See pages 524–527.)

4. Draw rhombus *K*(2, 2), *L*(4, 3), *M*(5, 5), *N*(3, 4). What translation is needed so that the image falls completely within quadrant III?

____

NAME

Daily Review 151

Trace each figure below and cut out the pattern. Use the pattern to determine if the figure tessellates. **Remember** that there can be no gaps or overlapping. (Need an example? See pages 538–539.)

1. 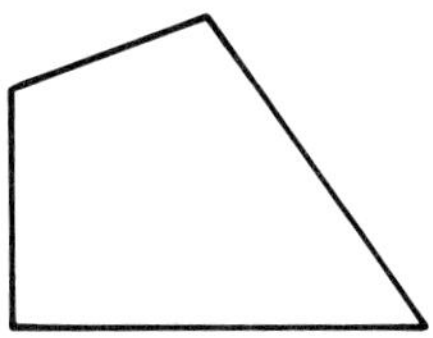

2. 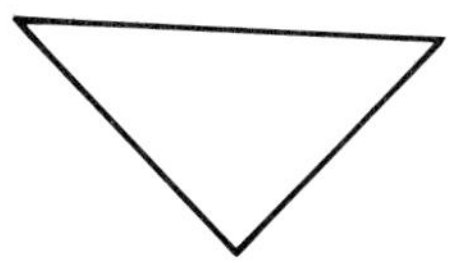

3. 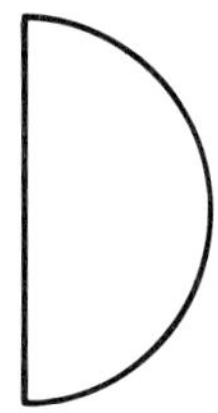

______________ ______________ ______________

Copy figure *EFG* onto your graph paper. Give the coordinates of the vertices of its image under each of the following rotations. (Need an example? See pages 532–535.)

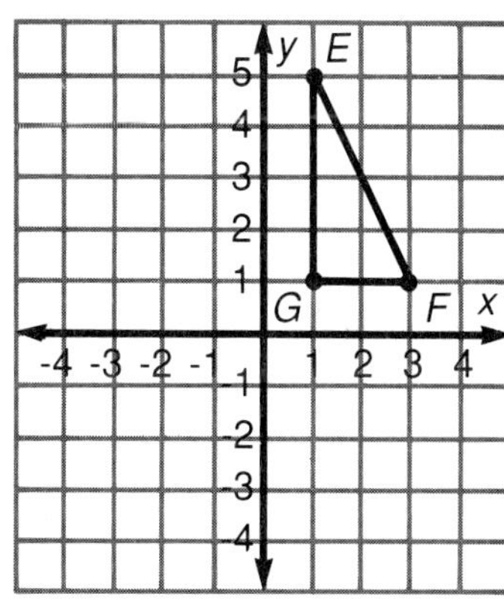

4. 90° counterclockwise rotation

____

5. 180° clockwise rotation

____

6. 90° clockwise rotation

____

NAME

Daily Review 152

Substitute the given values for x and find the corresponding values for y. (Need an example? See pages 540–541.)

1. $y = \frac{x}{5}$

x	y	(x, y)
−10		
−5		
0		
5		
10		
15		

2. $y = -4x - 1$

x	y	(x, y)
−2		
−1		
0		
1		
2		
3		

Trace each figure below and cut out the pattern. Use the pattern to determine if the figure tessellates. **Remember** that there can be no gaps or overlapping. (Need an example?. See pages 538–539.)

3.

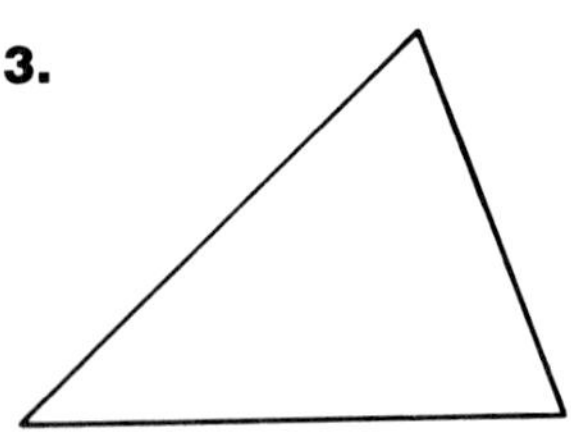

____

4.

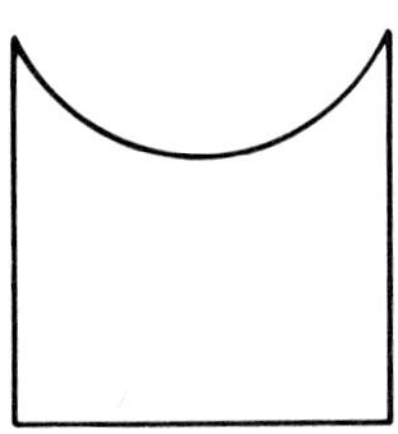

____

5.

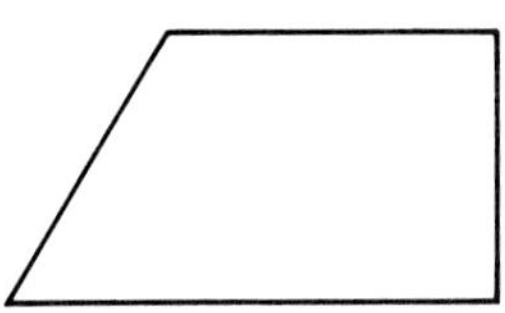

____

NAME ______________________

Daily Review 153

Select at least four values for x and find the corresponding values of y. Then graph the ordered pairs on another sheet of paper and connect the points. (Need an example? See pages 542–543.)

1. $y = x - 2$ **2.** $y = -x$ **3.** $y = 2x + 1$

4. $y = 3x$ **5.** $y = -4x + 3$ **6.** $y = -2x$

Set up a table to find ordered pairs to satisfy each equation. Solve for y first if necessary. (Need an example? See pages 540–541.)

7. $y = -2x - 2$

x	y	(x, y)

8. $x + y = 6$

x	y	(x, y)

Trace each figure below and cut out the pattern. Use the pattern to determine if the figure tessellates. **Remember** that there can be no gaps or overlapping. (Need an example? See pages 538–539.)

9.

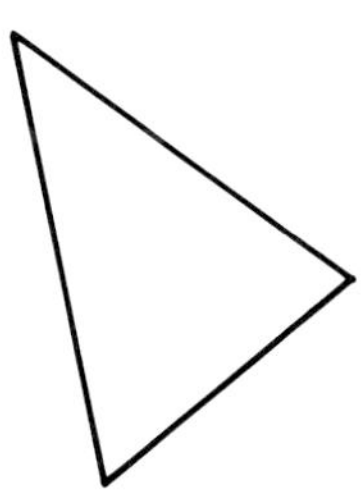

____

10.

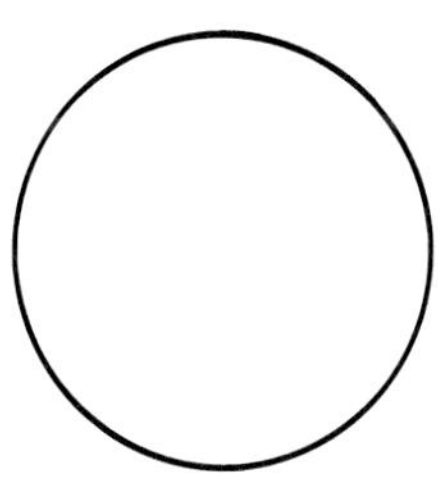

____

11.

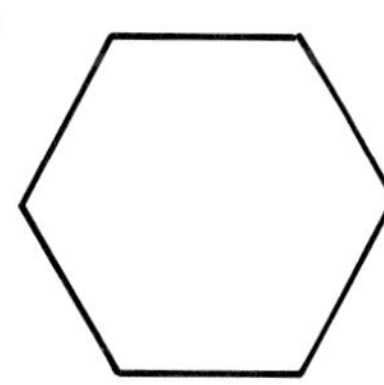

____

NAME

Daily Review 154

Graph each pair of equations on the same grid to find the solution. Check each solution in both equations. (Need an example? See pages 544–545.)

1. $y = -x + 5$
$y = 3x - 7$

______

2. $y = 2x$
$y = -x + 3$

______

3. $y = x + 4$
$y = -3x$

______

Select at least four values for x and find the corresponding values of y. Then graph the ordered pairs and connect the points. (Need an example? See pages 542–543.)

4. $y = 4x$

5. $y = -x + 1$

6. $y = -3x$

7. $y = x - 2$

8. $y = x + 3$

9. $y = 2x - 4$

For each of the following, set up a table to find ordered pairs to satisfy each equation. Solve for y first, if necessary. (Need an example? See pages 540–541.)

10. $3x + y = -2$

x	y	(x, y)

11. $y = \frac{x}{4} - 2$

x	y	(x, y)

NAME

Daily Review 155

Bill needs to rent a set of tools. Right Tools Inc. charges \$20 plus \$4 per hour. Tools-4-You Inc. charges \$30 plus \$3 per hour. (Need an example? See pages 546–547.)

1. Write the equation for the cost of renting tools from Right Tools.

____

2. What equation shows the cost of renting tools from Tools-4-You?

____

3. Bill needs the tools for a total of 11 hours. Which company is cheaper?

____

4. What is the difference in the amount charged by each company for 20 hours of tool rental?

____

Graph each pair of equations on the same grid to find the solution. Check each solution in both equations. (Need an example? See pages 544–545.)

5. $y = x + 2$
$y = 2x - 3$

____

6. $y = -2x$
$y = -x + 4$

____

7. $y = 2x - 1$
$y = 3x$

____

Select at least four values for x and find the corresponding values of y. Then graph the ordered pairs and connect the points. (Need an example? See pages 542–543.)

8. $y = x + 3$ **9.** $y = \frac{x}{3}$ **10.** $y = 2x + 3$

11. $y = -x + 4$ **12.** $y = -2x + 5$ **13.** $y = \frac{x}{2} - 3$